SpringerBriefs in Computer Science

For further volumes:
http://www.springer.com/series/10028

Siobhan Rockcastle · Marilyne Andersen

Annual Dynamics of Daylight Variability and Contrast

A Simulation-Based Approach to Quantifying Visual Effects in Architecture

Springer

Siobhan Rockcastle
ENAC-IA-LIPID
EPFL
Lausanne
Switzerland

Marilyne Andersen
ENAC-IA-LIPID
EPFL
Lausanne
Switzerland

ISSN 2191-5768 ISSN 2191-5776 (electronic)
ISBN 978-1-4471-5232-3 ISBN 978-1-4471-5233-0 (eBook)
DOI 10.1007/978-1-4471-5233-0
Springer London Heidelberg New York Dordrecht

Library of Congress Control Number: 2013939067

Printed on acid-free paper

Springer is part of Springer Science+Business Media (www.springer.com)

Preface

Daylight is a dynamic source of illumination in architectural space, creating diverse and ephemeral configurations of light and shadow within the built environment. It can generate contrasting levels of brightness between distinct geometries or it can highlight smooth gradients of texture and color within the visual field. Perceptual qualities of daylight, such as contrast and temporal variability, are essential to our understanding of both material and visual effects in architecture. With that in mind, how can architects measure the impacts of these dynamic and perceptual effects of daylight and compare them to other, task-based illumination and comfort metrics?

Under the rapidly growing context of energy conscious research, we need to rebalance our definition of "performance" to include those perceptual and aesthetic aspects of light that are often disregarded by the world of simulation. Contrast is important to the definition of space and it is essential in understanding how architecture is enhanced and transformed over time by the dynamic and variable characteristics of daylight. Although there are a growing number of studies that seek to define the relationship between brightness, contrast, and lighting quality, the dynamic role of daylight within the visual field is underrepresented by existing metrics. Although spatial contrast and light variability are fundamental to the visual experience of architecture, architects still rely primarily on intuition and experience to evaluate their designs, because there are few, if any, metrics that address these factors.

New metrics that address this challenge could help designers to contextualize the relative strength and temporal stability of contrast within a given architectural space, which would open up a new dimension in architectural performance. Through an analysis of contemporary architecture from around the world, we have developed a new typological language that categorizes architectural space in terms of contrast and temporal variation. This research proposes a new family of metrics that quantify the magnitude of contrast-based visual effects and time-based variation within daylit space through the use of time-segmented daylight renderings to provide a more holistic analysis of daylight performance.

Acknowledgments

The research for this book was conducted in partial fulfillment of the requirements for the Degree of Master of Science in Architecture Studies at the Massachusetts Institute of Technology in 2011. Since then, the research has been published in the proceedings to the simAUD conference in Orlando in 2012, where it received the 'Best Paper Award.' Since February of 2013, this research is being further developed in LIPID lab at the École Polytechnique Fédérale de Lausanne.

We would like to thank Professor Terry Knight and Professor Sheila Kennedy for their thoughtful contributions to this research.

Contents

Chapter 1
Introduction

Keywords Daylight architecture · Architectural typologies · Spatial definition · Contrast · Luminous diversity

1.1 Visual Perception in Daylight Architecture

> A building speaks through the silence of perception orchestrated by light. Luminosity is as integral to its spatial experience as porosity is integral to urban experience. (Holl, 2006)

Most architects would agree that daylight is an important asset to the design of good architecture, but *what* aspects of natural light quantify or qualify the visual performance of a space? Perceptual qualities such as contrast and temporal variability are essential to our appreciation of architectural space; natural illumination adds depth to complex geometries and infuses otherwise static interior spaces with shifting compositions of light and shadow. And while architecture is greatly altered by the ephemeral and perceptual qualities of daylight, there is a lack of metrics that address these factors on a dynamic scale.

In today's context of heightened environmental awareness, we feel pressure to evaluate architecture in terms of sustainable performance criteria. As designers, we are trained to place value in the concept of spatial experience; however, we are increasingly asked to quantify our design intentions in terms of net energy balance. As these requirements become more pervasive in our architectural education and the justification of design quality, we must position the term 'environmental' to include those perceptual qualities of light that have become secondary in our dialogue about performance. Architecture must 'perform' in both qualitative and quantitative criteria, and we must work to re-establish the role of perceptual and preferential indicators in our language about performance. Architects choreograph light to enhance the perception of space and draw attention toward elements of visual significance. 'Light reveals architecture, and in return, architecture must reveal light (Millet 1996).'

S. Rockcastle and M. Andersen, *Annual Dynamics of Daylight Variability and Contrast*, SpringerBriefs in Computer Science,
DOI: 10.1007/978-1-4471-5233-0_1,

The very character and purpose of light is dependent on a set of design principles which are revealed to the observer through experience, and not through a planar map of illumination levels. We may ask ourselves, what does begin to distinguish these varied characteristics of light and how might we develop an understanding of their perceptual effects in architecture? How does daylight vary from one location to the next and how do hourly and seasonal changes in quantity and orientation alter its visual impacts within space?

1.2 The Ephemerality of Natural Light

Unlike artificial light sources, which can be adjusted to meet a desired visual effect regardless of location and time, daylight is sensitive to an array of influences. The latitude of a given location affects the length and intensity of daylight hours throughout the year, while local changes in climate affect its hourly strength and variability. Surrounding site conditions can amplify or diminish the sun's ability to penetrate an interior space and it is often difficult to predict how these conditions will change over time, especially within the complex fabric of an urban environment.

> As light passes through small holes, it spreads out, frays and bends. The resulting shadows do not necessarily look like the silhouettes of the objects that cast them. Light bends in ways that yield shadows with bright bands, dark bands, or no sharp edges. (Holl, 2006)

How then, can we inform architecture with a richer understanding of this dynamic and variable source of illumination so that we can incorporate its perceptual effects alongside energy and comfort-related design criteria?

In their book titled *Environmental Diversity in Architecture*, Mary Anne Steane and Koen Steemers discuss the importance of environmental and visual diversity in the built environment, describing the need for both temporal and spatial diversity in architecture. Steane describes a number of ways in which a building can encourage temporal diversity through its orientation, the size and location of its apertures, and the spectral quality of its finishes. In a study conducted on the relationship between luminance diversity and the perceived quality of interior space, the more diverse the luminance in the field of view, the more pleasant and visually warm the space was reported to appear (Steane and Steemers 2004). The same study reported that students in a library were turning on individual task lights even though illuminance levels measured well above an acceptable level at the work plane (Steane and Steemers 2004). It was inferred that the student's desire for more light was not related to inadequate illuminance levels, but to a *desire for diversity* within their visual field. This raises an important issue in the discussion on daylight analysis in architecture. Although many of our codes and recommendations are concerned with task-based illumination levels, occupants are attracted to the visual diversity of their surroundings, establishing the need for new metrics that can quantify and place value in these perceptual qualities.

1.3 Defining the Value of Light in Spatial Definition

In order to understand the perceptual effects of daylight in architecture, it is necessary to define the role of contrast in spatial definition. Spatial definition depends on the balance between light and dark, the eye's ability to perceive those differences, and the brains ability to translate that information into an understanding of depth and complexity (Liljefors 1997). In a sense, this notion of space is entirely dependent on the photo sensors in the human eye and the brain's interpretation of that information into a kind of map. While illumination levels determine whether we can see our surroundings, contrast and brightness determine the complexity and richness of its perceived composition. The luminous effect can be described as a combination of four factors: the source (its intensity, its directional characteristics, its color); the geometry (its relationship between source and receiving surfaces); the surfaces that receive or modify light (becoming secondary light sources in themselves); and the person who views the source and illuminated surfaces as he or she moves around (Millet 1996).

The evaluation of these four elements into a universally applicable set of preferences or design criteria is not, however, a simple task. We can experience pleasure in a diverse mix of spaces that represent both high and low contrast, dynamic and static lighting conditions. The human brain is subjective in its response to formal composition and the use of light and contrast in the disciplines of art and architectural design is varied. If we want to develop a strategy for quantifying and/or comparing any of these luminous effects, there is the further challenge of documenting light within a static image such as a painting, rendering, or photograph. Although high dynamic range HDR cameras can now utilize multiple exposures to more accurately capture a photograph that mimics the human eye (Ward 1994), the struggle to represent light in a photo-realistic manner is a challenge that has evaded the field of representation for centuries.

In the seventeenth century, the Dutch painter Johann Vermeer was known for his ability to render light and color with a richness that surpassed his contemporaries. In his painting entitled Young Woman with a Water Pitcher (Fig. 1.1), Vermeer captured the tonal variations in light as they was filtered by the stained glass window and absorbed by the fabric and skin of the female subject. What was most impressive about Vermeer's work was the way in which he could capture diffuse light as it was transmitted through or bounced off of objects in the surrounding scene. His paintings came alive through the thin and arduous layering of pigments which describe the tonal complexity of each surface (Alpers 1983). In the eighteenth century, the Italian painter Antonio Canaletto pushed the perception of spatial depth through the blurring of objects located farthest away from the foreground of the visual field and the projection of shadows out into the perspectival view. Using a camera obscura to simulate the depth of a given scene, he was able to more accurately render the effects of illumination and detail as it would be experienced from the perspective of an observer (Canaletto, reprinted in 1971).

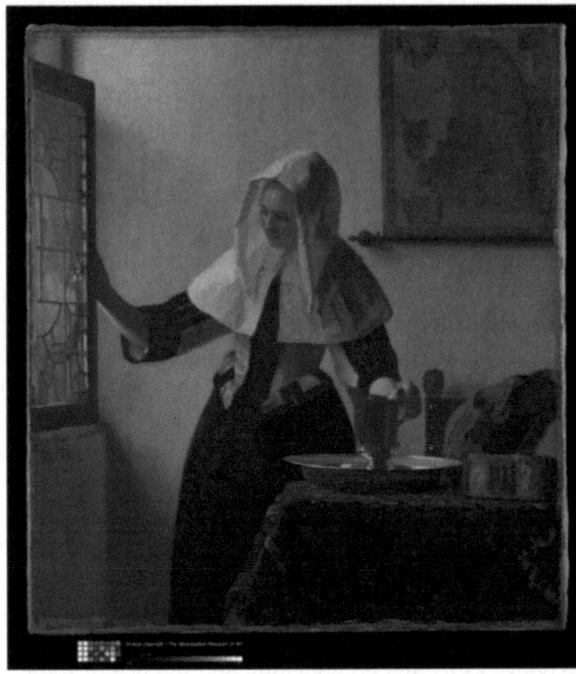

Fig. 1.1 Young woman with a water pitcher. The Metropolitan Museum of Art, marquand collection, gift of Henry G. Marquand, 1889 (89.15.21) © The Metropolitan Museum of Art

The difficulty in accurately representing light and its perceived visual effects continue to challenge architects and daylighting designers today. We still struggle with the accuracy and time intensive nature of rendering light as well as our methods for describing and calculating the quantitative and qualitative nature of that light. As the techniques of painting continued to evolve toward more realistic methods of light rendering and spatial representation in the nineteenth century, artists in the twentieth century began to unpack the notion of space as a compositional map of color and contrast. The work of Piet Mondrian represents this departure from object and field to an abstracted two-dimensional space (Ching et al. 2011). Mondrian's evolution from an impressionistic style to a more abstract and orthographic interpretation of space can be seen through Red Tree Oil on canvas (1908), Composition II (1930) and Composition IX (1939–1942).

The architecture of the modern movement followed this same trend as architectural expression began to move away from the voluptuous and ornamental toward a more functional machine esthetic. If we discuss the architectural intentions of seventeenth century Baroque architecture with those of twentieth century Modernism, we can see a dramatic shift in the expression of volume and the choreography of light. Baroque architecture embraced the volumetric massing of bold elements and curved domes, employing light as a figure that emphasized the geometry of space (Ching et al. 2011). This expression can be seen in Francesco Borromini's San alle Quattro Fontane in Rome (Fig. 1.2). Modern architecture,

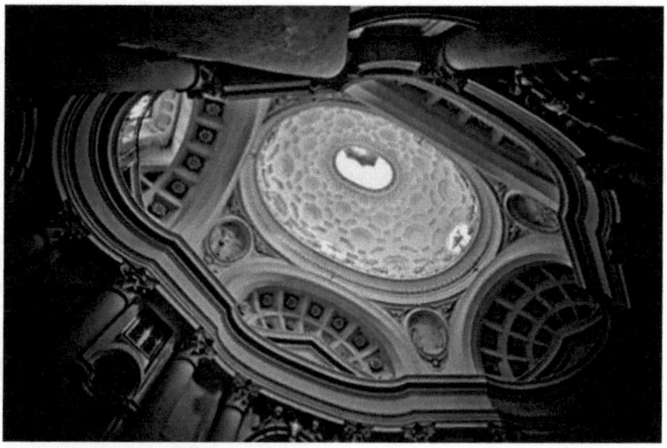

Fig. 1.2 San Carlo alle Quattro Fontane batintherain, 'God in a nutshell' December 31, 2008 via flickr, creative commons license

however, stripped classical ornamental stimuli and drew attention to the ordered composition and functional expression of space. The Barcelona Pavilion, designed by Mies van der Rohe and completed in 1929, exemplifies these qualities (Fig. 1.3). An embrace of transparency and new advances in materials and technologies developed alongside this reduction in ornament, liberating architecture from an adherence to the orders of past generations (Ching et al. 2011; Curtis 1996).

We can reflect upon the shifting forces that have impacted architectural history, but the fact remains that human preference; toward spatial definition, material form, and light, is subjective. Perhaps the one thing we do know is that luminosity, contrast, and their role in defining space is a highly charged topic in architectural expression. In the last two decades, we have experienced an emergence of more complex surface geometry and a renewed sense of delight in the interaction between elements of the natural and built environments. Categories of architectural form have grown increasingly more diverse as geometric modeling software has liberated the architect from a dependency on flat or regular surfaces and modes of fabrication. The result of this liberation includes some highly dramatic and articulated spaces whose interaction with direct sunlight brings the question of contrast visual perception to the foreground of any discussion on daylighting design.

While some spaces are designed for task-oriented activities (i.e., classrooms, art studios, and/or galleries) and require specific illumination levels to perform visual tasks, many do not require this level of control should not be subjected to the same performance criteria. Task-driven illumination and comfort metrics must be considered alongside perceptual performance metrics to ensure that a more holistic set of design goals is supported and achieved. In addition to holistic performance goals, architects must learn to assess the dynamic impacts of luminosity throughout *space* and *time* to achieve a stronger link between energy, comfort, and perceptual performance.

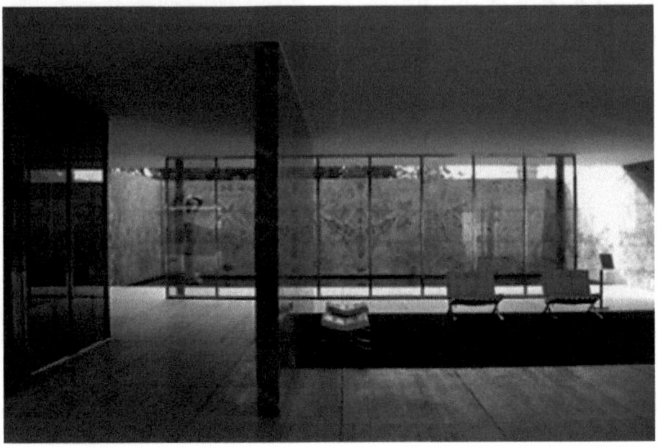

Fig. 1.3 Barcelona Pavilion Harshil.Shah (Harshil Shah), 'Barcelona—Pavelló Mies van der Rohe' June 7, 2008, via flickr, creative commons license

1.4 Typological Approaches to Daylight Design

Visual interest in architectural daylighting could be described as the esthetic and perceptual aspects of illumination that render a space *interesting*. The subjective nature of design makes indicators such as visual interest difficult to define, but a closer look at contemporary architecture from around the world suggests that there are certain similarities in how architects choose to choreograph daylight for varied programmatic needs and experiential effects. These *types* of daylight could be organized into a series of strategies that can foster a language about the qualitative effects of illumination in architectural space. For example, the direct and dramatic penetration of sunlight through the Kogod Courtyard at the Smithsonian Institute highlights the intended ephemerality of its use (Fig. 1.4). The courtyard is intended for occasional occupation by its visitors who are moving through the space en route from one location to another. They have no need for controlled illumination levels or protection from direct sunlight. Some may argue that this fleeting connection to the harsh perceptual effects of light and shadow evokes a certain delight

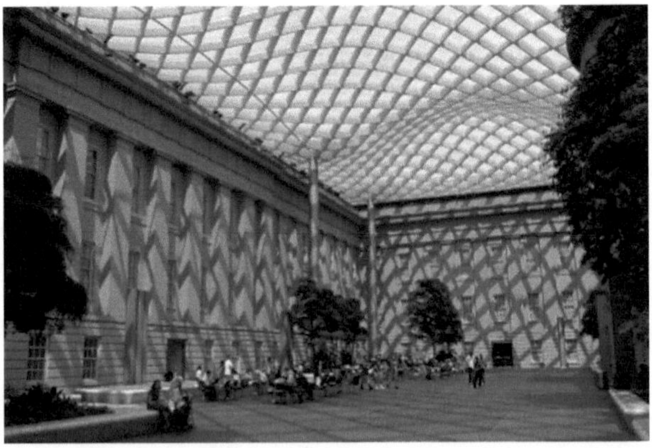

Fig. 1.4 Kogod Courtyard AgnosticPreachersKid, 'The Kogod Courtyard' May 29, 2010 via wikimedia commons, creative commons license

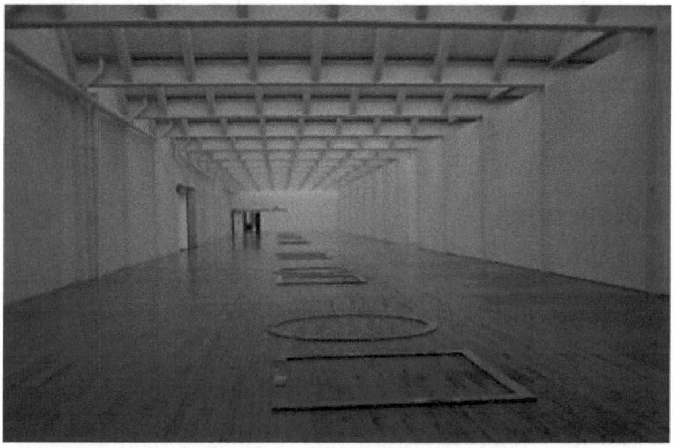

Fig. 1.5 Dia Beacon Museum Yusunkwon, August 20, 2004 via flickr, creative commons license

in the human subject, who spends the rest of his day trapped within the monotony of his office cubicle (Steane and Steemers 2004).

On the opposite side of the spectrum, the north-facing monitors that illuminate the Dia Beacon Museum (Fig. 1.5) in upstate New York cast an even and unfaltering light onto the tightly acclimatized environment of the galleries. These spaces were designed to maintain an even distribution of daylight without drawing attention away from the artwork or the scale and uniformity of the appropriated warehouse. In this case, contrast and light variability are kept at a minimum to achieve the intended spatial effects of the architectural design.

Through an analysis of these spaces and others, it becomes clear that we need new daylight performance criteria that can address a more diverse range of programmatic uses and perceptual design goals. A comprehensive study of contemporary global architecture will allow us to categorize interior spaces according to their daylight design strategy and resulting visual effect. We can then take a critical look at existing daylight performance metrics through the lens of these architectural examples to identify the aspects of illumination that are *not* being thoroughly evaluated. If existing illumination and visual comfort metrics for task performance evaluate one dimension of lighting performance, then this research will strive to unearth those alternate dimensions and develop a vocabulary of daylight-driven effects that further our understanding of perceptual performance in architecture.

This research will introduce the need for visually dynamic metrics through a critical analysis of existing daylight performance tools in the context of contemporary architecture. Through a survey of existing spaces, this research will develop a new typological approach for measuring spatial and temporal diversity in daylight architecture. Using this typological study, we will propose three new metrics for describing and quantifying contrast and temporal diversity through the medium of digital images. These metrics will then be applied to a series of case study spaces to pre-validate their success in quantifying those qualitative visual effects unearthed in the typological study. In the final chapter, these metrics will be applied to a series of existing architectural spaces and compared against current daylight performance metrics to discuss the need for a more objective and holistic approach to daylight analysis.

References

Alpers, S. (1983). *The art of describing: Dutch art in the Seventeenth Century*. Chicago: University of Chicago Press.

Canaletto, A. (reprinted in 1971). Views of venice. New York: Dover Publications.

Ching, F., Jarzombek, M., & Vikramaditya, P. (2011). *A global history of architecture*. Hoboken: Wiley.

Curtis, W. (1996). *Modern architecture since 1900*. London: Phaidon Press.

Holl, S. (2006). *Luminosity/porosity*. Tokyo: Toto.

Liljefors, A. (1997). Lighting and color terminology. *Paper Presented at a CIE Discussion*. Stockholm: Comission Internationale de l'Eclairage.

Millet, M. (1996). *Light revealing architecture*. New York: Van Nostrand Reinhold.

Steane, M. A., & Steemers, K. (2004). *Environmental diversity in architecture*. New York: Spoon Press.

Ward, G. (1994). The RADIANCE lighting simulation and rendering system. *Proceedings of '94 SIGGRAPH Conference*, (pp. 459–472).

Chapter 2
Research Context

Keywords Daylight performance metrics · Task-based illumination · Visual comfort for task performance · Contrast · Luminous diversity

Of the many established metrics that quantify daylight performance, a disproportionately small group of these address factors of perceptual appeal. An obvious reason for this is that most metrics were developed to improve energy efficiency by replacing electric lighting, or to avoid human *discomfort* due to sources of glare within the visual field. Although architects use sunlight to choreograph the perceptual quality of space, there is limited research available to help designers understand the complex variability of daylight across an occupant's visual field. While there is some agreement on the minimum amount of illumination that is required for the human eye to perform visual tasks within a given space, there is little consensus on how much contrast or brightness makes a space visually appealing. Those studies that do address the luminous field-of-view are limited in their analysis of contrast composition and do not address the temporal variation that occurs due to the daily and seasonal variations in solar orientation.

Through a comparison of existing interior spaces, this chapter will introduce a range of daylight design strategies found in global contemporary architecture. Each strategy varies in its approach to sunlight penetration and daylight distribution, yet reinforces a specific spatial experience that is central to the architectural goals of the project. It is through these architectural spaces that we will introduce the role of contrast and temporal diversity as an indicator of visual design performance and discuss the need for new perception-driven metrics to complement existing *task-driven* and *comfort-based* performance metrics. Within the field of architecture, it is essential that we couple daylight performance criteria with design intent and provide metrics that address visual, perceptual, and task-related criteria.

S. Rockcastle and M. Andersen, *Annual Dynamics of Daylight Variability and Contrast*, SpringerBriefs in Computer Science,
DOI: 10.1007/978-1-4471-5233-0_2, © The Author(s) 2013

2.1 Contrast as an Indicator of Qualitative Performance

In architecture, spatial definition depends on the balance between light and dark, the eye's ability to perceive those differences, and the brain's ability to use that information to understand the depth and complexity of our surroundings. To introduce the importance of contrast in architecture, we will look at four contemporary examples and examine the differences inherent in their expression of contrast and spatial differentiation.

The first example is Norman Foster's renovation of the Kogod Courtyard in Washington, DC (Fig. 2.1). The articulated glass roof structure of the courtyard allows for a dramatic penetration of direct sunlight, imposing strong patterns of contrast onto the walls and floor of the interior space. Designed for temporary occupation and public gathering, the space's programmatic use does not require a tightly controlled lighting strategy. On the contrary, it takes advantage of the dynamic nature of sunlight through transparency to create a diverse and visually engaging environment for its occupants.

The second example, Herzog and De Meuron's Dominus Winery located in Yountville, California (Fig. 2.2), differs in its attitude toward the surrounding environment, allowing light to filter in through an exterior gabion wall. The architects sought to create a unified relationship to the landscape, using local stones to provide a naturally cool thermal environment with visually engaging effects. The interior spaces maintain a variable relationship to incoming light, but the overall lighting levels are dim in comparison with the Smithsonian Courtyard. Occasional spots of direct sunlight on the floors and walls of the circulation corridor create an abruptly contrasted environment. This daylight strategy filters direct sunlight from the south-facing façade while drawing attention to the materiality of its exterior wall, highlighting the seemingly organic non-uniformity of its composition (Ursprung 2002). One could argue that this strategy produces a highly contrasted interior like that of the Smithsonian Courtyard, but with more controlled variations over the course of the day and a darker base composition, overall.

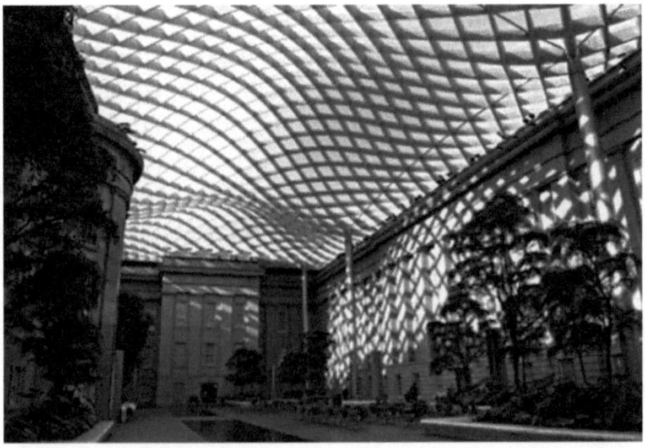

Fig. 2.1 Kogod Courtyard dctim1, 'Kogod Courtyard—northeast corner and floor—Smithsonian American Art Museum' January 04, 2013, via flickr, creative commons license

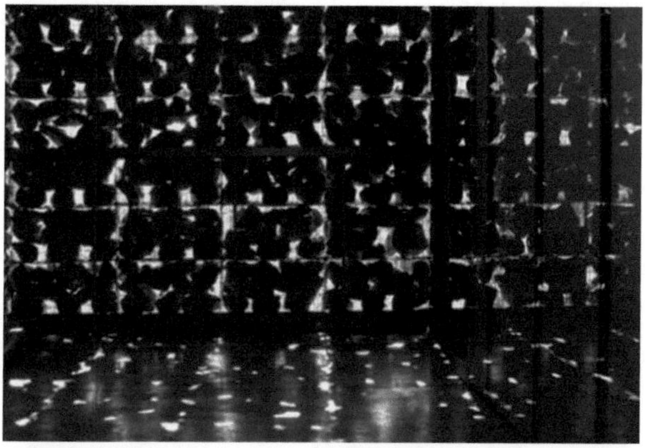

Fig. 2.2 Dominus winery © Dominus Estate, Yountville, CA, USA

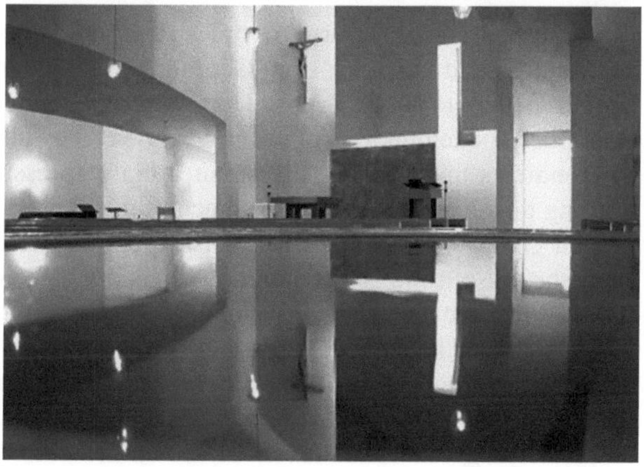

Fig. 2.3 Church of St. Ignatius Joe Mabel, 'Chapel of St. Ignatius' November 30, 2007, via wikimedia, creative commons license

For the third example, we will consider Steven Holl's Church of St. Ignatius in Seattle, Washington (Fig. 2.3). This space is vastly different in character from the two previous examples, composing sunlight into a series of carved, indirect figures which accentuate its volumetric qualities (Holl 1999). The light within this church could be described as more selectively diffuse, with compositional lines and volumes being defined through distinct spatial geometries. This example represents less extreme contrast than that of the Smithsonian Courtyard or the Dominus Winery, but still maintains a dynamic relationship to the exterior as shifting light levels cause figural volumes of light to change over time.

The final example, Renzo Piano's High Museum of Art in Atlanta, Georgia (Fig. 2.4), employs an indirect daylighting strategy similar to that of the Church of St. Ignatius. However, it differs in the stability of its internal illumination as the light tubes that compose the roof collect and distribute diffuse light from the north. The programmatic use of this space as a gallery necessitates an even distribution of internal lighting levels while preventing any direct sunlight that may cause damage to or distract from the artwork. As a result, the presence of strong contrast and temporal instability is minimized across the space.

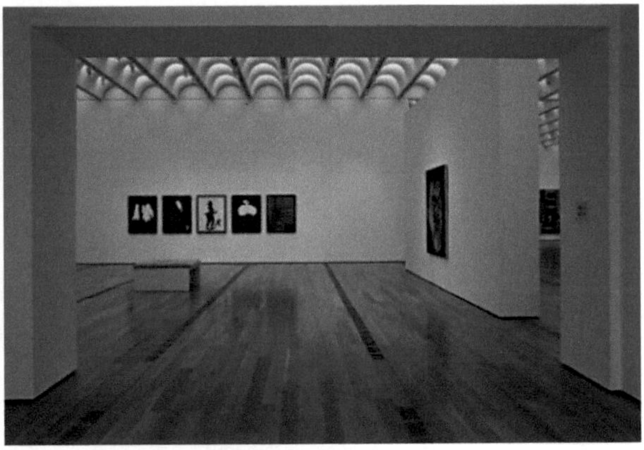

Fig. 2.4 High Museum Brookenovak, 'Wandering around the High' September 8, 2007, via wikimedia, creative commons license

These four contemporary examples represent varied site conditions, both urban and rural; varied latitudes, from Georgia to Seattle; and varied programmatic uses from art gallery to public atrium. They represent dramatically different compositions of contrast and temporal light stability, and yet they all produce visually stimulating environments that enhance the architectural expression of interior space. In considering this diverse range of architectural examples, our goal is to define the perceptual characteristics that distinguish them and determine what this can tell us about the role of contrast and luminous diversity in the visual performance of interior space. While the notion of perceptual 'quality' is, admittedly, a difficult element to quantify due to its subjective nature, we believe that there are metrics that could measure the compositional impacts of contrast and luminance diversity and help inform architects about their varied effects over time. Although we have no intention of prescribing universal threshold recommendations for contrast or luminance diversity, we feel that establishing a method for quantifying these compositional effects will provide architects a tool for comparing design options and contextualizing those options within a relative scale. Through measuring and comparing the impacts of spatial contrast and luminance diversity over time, architects will be able to communicate their objectives more comprehensively and choreograph the dynamic visual effects of a space to meet their intended design goals. In turn, this relative scale will serve as a foundation for new dynamic design metrics that measure spatial contrast and luminance diversity in daylight architectural space.

2.2 Spatial Considerations for Daylight Performance

Using these examples as context, we will now transition into a critical analysis of existing daylighting performance metrics to build a case for more visually dynamic methods as they relate to spatial contrast and daylight variability. Existing daylight performance metrics can be divided into three main categories: illumination for task-driven performance, visual comfort for task-driven performance, and occupant preference toward the field-of-view. The methods explored in this research do not seek to discount existing metrics, but rather to contribute to a more holistic definition of performance. To achieve high-performance architecture, we must consider existing task-driven and visual comfort metrics along with new methods for evaluating temporal visual performance, in order to reaffirm the importance of perceptual factors in daylighting design.

2.2.1 Illumination for Task Performance

Before we can discuss those metrics that define daylighting performance within a building, it is important that we define the units of measurement used to quantify light. Illuminance, which describes the total luminous flux that falls on a surface, per unit area (CIE 1926), is the most widely applied measurement of light and is the foundation upon which most other task-driven metrics such as daylight factor and daylight autonomy are based. Codes and standards most commonly reference illuminance measurements across a work plane to determine the amount of light recommended for various tasks (IESNA 2000). Most task-based illuminance metrics were developed to analyze minimum threshold levels in task-oriented spaces such as offices, libraries, and schools (Lam 1977), and while these thresholds can be seen as somewhat subjective, they were established to ensure that adequate illumination could be measured and achieved across a given task surface for a given activity (IESNA 2000).

As far as practice and standards are concerned, daylight factor (DF), which measures the ratio between indoor and outdoor illuminance under overcast sky conditions (Moon and Spencer 1942), may be the most ubiquitous task-based illuminance metric in use (Fig. 2.5). This metric was originally created to estimate daylight access from a 'worst-case' perspective (Reinhart et al. 2006) while avoiding

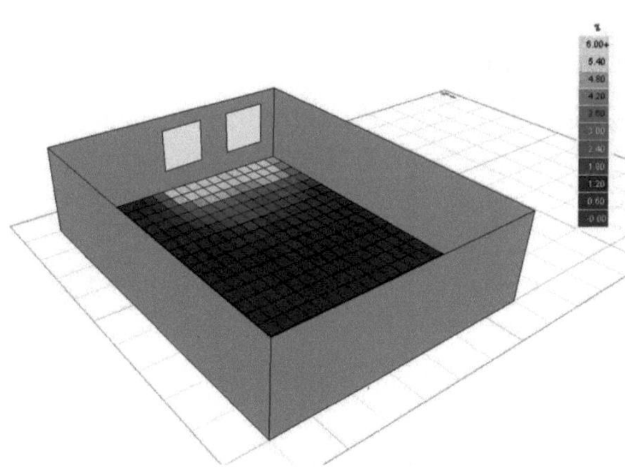

Fig. 2.5 Daylight factor simulation in ECOTECT, http://usa.autodesk.com/ecotect-analysis/

the difficulties associated with fluctuating sky conditions and the dynamic nature of sunlight (Waldram 1950). From an architectural standpoint, however, DF limits our understanding of daylight as a dynamic source of illumination, assuming a 'more-is-better' attitude, regardless of sky type (direct sun versus diffuse sky), climate, or programmatic use of the space under consideration (Reinhart et al. 2006).

If we were solely concerned with bringing light into a building, then we could maximize our lighting scheme using DF, but many of the problems we face in architectural design deal with controlling, animating, and understanding the impacts of direct sunlight under varied conditions (Steane and Steemers 2004). In the case of the High Museum by Renzo Piano, the use of DF would provide little value to the optimization of its daylighting strategy, which seeks to control the penetration of direct sunlight. Likewise, the DF is hardly an effective guide for the design of spaces like the Dominus Winery, by Herzog and deMeuron, where high-contrast, low-light conditions are preferred.

Over the past few decades, there have been significant improvements in our understanding of daylight as a dynamic source of interior illumination. We have transitioned from static metrics such as DF to annual climate-based metrics such as daylight autonomy (DA) (Reinhart et al. 2006) and useful daylight illuminance (UDI) (Nabil and Mardaljevic 2006), and goal-based metrics such as acceptable illuminance extent (AIE) (Kleindienst and Andersen 2012) to account for a more statistically accurate method of quantifying internal illuminance levels (Mardaljevic 2000).

Daylight autonomy (DA) was first defined as the percentage of a year when the minimum illuminance threshold was met by daylight alone and did not require supplemental electric lighting. In 2001, it was redefined as the percentage of *occupied time* throughout the year when the minimum illuminance requirements at a sensor are met by daylight alone (Reinhart and Walkenhorst 2001). As a metric, DA can evaluate annual illuminance thresholds, taking into account building orientation and climate-driven sky types. It is useful in determining whether a surface within a space achieves a minimum threshold of illuminance and what part of the year that threshold is maintained (Fig. 2.6).

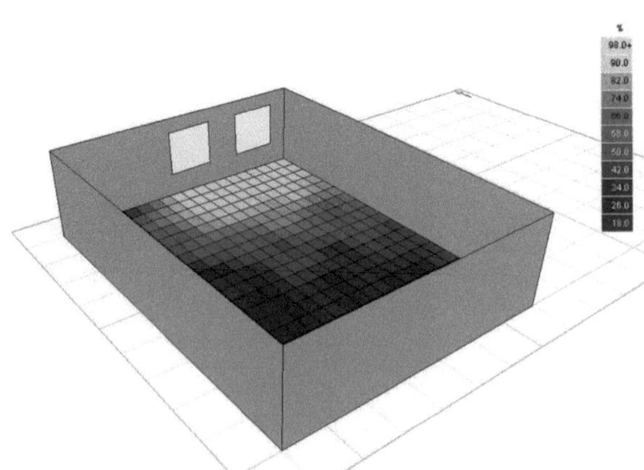

Fig. 2.6 Daylight autonomy ECOTECT, http://usa.autodesk.com/ecotect-analysis/

Continuous daylight autonomy (DAcon) is a similar method of evaluating annual performance through illuminance thresholds across a sensor plane. It awards partial credit for illuminance levels that fall below the minimum threshold on a weighted scale, supporting the notion that some daylight is still better than no daylight (Rogers 2006). This approach allows for a smoother gradient of threshold compliance, accommodating research which concluded that many people work comfortably at illuminance levels below standard minimum thresholds of 500 or even 300 lux (Reinhart and Voss 2003).

2.2.2 *Visual Comfort for Task Performance*

Unlike task-based illumination metrics that rely on illuminance, successful task-based visual comfort metrics (typically pertaining to glare) rely on luminance, defined as the amount of light emitted by a surface in a given direction (CIE 1926). Of the four photometric quantities (flux, intensity, illuminance, and luminance), luminance is closest to how the eye perceives light and, as such, appears to be the only quantity capable of expressing visual discomfort.

As luminance, brightness, and contrast are subjectively evaluated, glare discomfort is fragmented across no less than seven established metrics (Wienold and Christoffersen 2006; IESNA 2000; Osterhaus 2005). Daylight glare probability (DGP) (Wienold and Christoffersen 2006), considered the most reliable index for side-lit office spaces, is the only index that relies on daylighting conditions. While these indices do not always agree, partly due to the fact that some were developed for electric lighting sources and others for daylight, most are derived from the same four quantities: luminance, size of the glare source, position of the glare source, and the surrounding field of luminance that the eye must adapt to (Wienold 2009).

Daylight glare probability (DGP) is the percentage of people that are disturbed by daylight-based sources of glare in a side-lit office environment (Wienold and Christoffersen 2006). The resulting value, a percentage between 0 and 100, has only been validated for 20 % DGP or higher. Like other glare indices, DGP too was developed for task-oriented settings (Kleindienst and Andersen 2012). Comfort-based metrics such as DGP must be used carefully, as many architectural spaces do not require low-glare tolerance in their programmatic use and some even celebrate high contrast as an intentional visual effect. Figure 2.7 shows an example DGP analysis produced using the DIVA toolbar (http://www.diva-for-rhino.com, 2009), an analysis plug-in developed for Rhinoceros 4.0 (http://www.rhino3d.com, 2007) by the Harvard Graduate School of Design.

An annual DGP analysis (one rendering for every hour of available sunlight) using common RADIANCE rendering routines and *evalglare* requires substantial computing time. A simplified method, known as DGPs, was developed to minimize computational intensity while providing a reasonable assessment of side-lit office spaces where direct sun transmission does not impact the observer (Wienold 2009). To further explore the *dynamic* assessment of glare within a standard work

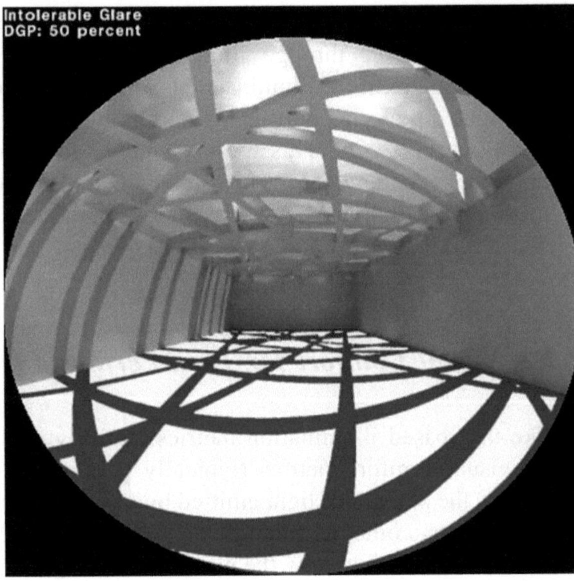

Fig. 2.7 Daylight glare probability, DIVA for rhinoceros, http://www.diva-for-rhino.com/

environment, the concept of an 'adaptive zone,' which accounts for occupant freedom to change position and view direction, was tested across five glare indices (Jakubiec and Reinhart 2012). DGP was found to be the most robust and accurate metric of those tested, while the enhanced simplified DGP method (Wienold 2009) was found to produce a comprehensive yearly analysis with a reasonable amount of computing power (Jakubiec and Reinhart 2012).

2.2.3 Evaluating the Perceptual Field-of-View

While comfort-based luminance metrics such as DGP extend our quantitative methods of assessment beyond task-based illumination metrics such as DF and DA, the current state of lighting research is still generally dominated by what Cuttle would refer to as the rut of a nineteenth-century concept (Cuttle 2010). Lighting research has been historically dominated by task-performance and visual comfort criteria, which are only applicable to spaces where visual tasks are frequently encountered. For spaces where visual task performance is less indicative of lighting performance, we often seek to create acceptably bright and/or visually engaging environments (Cuttle 2010). To evaluate occupant *satisfaction* with the perceptual field-of-view and measure the *positive* impacts of luminosity within interior architecture, past research has relied on measurements such as average luminance, threshold luminance, and luminance diversity in line with occupant surveys to establish trends in preference.

Two dimensions that are widely accepted to impact the field-of-view are average luminance and luminance variation (Veitch and Newsham 2000). The former has been directly associated with perceived brightness and the latter with visual interest (Loe et al. 1994). As brightness is subjectively evaluated by the human brain, contrast and luminous composition are often regarded as qualitative indicators of daylight performance, prompting researchers to use empirical methods (i.e., surveys) to establish a relationship with occupant preference.

While renderings and digital photographs are used by architects to communicate design intent, high-dynamic range (HDR) images produced through RADIANCE can provide an expanded range of photometric information, allowing us to gain luminance values and evaluate characteristics such as brightness and contrast (Ward 1994).

In a study conducted by Cetegen et al. occupant surveys were used to establish a direct correlation between the average luminance across an HDR image and its perceived 'pleasantness' or relative 'excitement' (Cetegen et al. 2008). In this study, participants were shown digital HDR images of an office environment with varying partition configurations and view conditions. For each of the configurations, the participants ranked the images in terms of their satisfaction with the amount of view, light, and their own visual comfort. The results found a positive trend between increased average luminance levels and satisfaction for the view as well as increased luminance diversity and the participant's impression of excitement (Cetegen et al. 2008). It was determined that both *average luminance* and *luminance diversity* contributed to occupant preference.

In an experiment conducted by Tiller and Veitch, participants were asked to adjust the brightness between two offices (using a dimmer switch) until they reached a perceived equilibrium in brightness; one office had a uniform lighting distribution and while the other had a non-uniform lighting distribution. Both offices had the same average luminance across the observed field-of-view. Task-plane illuminances were taken in each space, and it was determined that the office with a non-uniform luminance distribution required 5–10 % less work-plane illuminance to achieve the same level of perceived brightness as the office with a uniform lighting distribution (Tiller and Veitch 1995). The researchers concluded that *luminance distribution* across an occupant's field-of-view does, indeed, impact the perception of brightness within a given space.

In a study on visual comfort, participants were asked to adjust a set of horizontal blinds within a side-lit office space until the light distribution reached a level that they felt was 'most preferable,' and then again into a position that they felt was 'just disturbing' (Wymelenberg and Inanici 2009). HDR photographs were taken after each adjustment and used to run a series of luminance metrics to analyze the participant's selection of scenes. While the results established an upper threshold value over which the average luminance of the office was considered disturbing by all participants, the study was unable to determine a lower threshold given the diversity of results. DGP was calculated for each selected scene, but there were no significant trends between the 'most preferable' and 'just disturbing' spaces. The best predictive metrics for occupant preference in this study were found to be predetermined luminance threshold values (Lee et al. 2007) and standard deviation of luminance values. The authors concluded that *adequate* variations in luminance created a stimulating visual environment, while *excessive* luminance variability tended to create uncomfortable spaces (Wymelenberg and Inanici 2009).

A study of particular relevance to this research established a new method for measuring luminance diversity, called the Luminance Differences (LD) index (Parpairi et al. 2002). While efforts to use standard deviation to predict occupant discomfort have had some success, predicting positive preferences toward luminance diversity has been less successful. This is because the previous studies were unable to quantify local variations and thus identify patterns that would trigger visual interest. LD is calculated by taking eye-level luminance measurements in a 360° polar array across a horizontal plane and then calculating the difference in luminance levels across a range of acceptance angles corresponding to eye and head movement (Parpairi et al. 2002). LD allows us to calculate the perceived *noise* or variation in luminance values across our field-of-view. In this study, participants were asked to answer a questionnaire on their impressions of three Cambridge libraries across a series of predetermined viewpoints. LDs were calculated for each view position and then compared against the surveyed data to draw conclusions about luminance diversity and occupant preference. The authors concluded that luminance variability was highly appreciated by the subjects in all three library spaces and that the more variable the luminance across the field-of-view, the more 'Pleasant', the spaces were perceived to be. Furthermore, high luminances were not required to achieve satisfaction—variability was found to contribute more to occupant satisfaction than power.

The studies discussed so far rely on occupant surveys as an empirical method for measuring human preferences toward luminosity within the perceptual field-of-view. Another category of research focuses on the analysis of architecture to measure the *relative* performance of light between existing spaces. An example of this research can be seen in Claude Demers' daylight classification system (Demers 2007). In her work on contrast and brightness analysis through the use of digital images, Demers used grayscale histograms to identify the dominance of bright, dark, and middle-range pixel values within interior architecture. Based on the mean brightness (average luminance) and standard deviation of those pixel values, she has developed a daylight classification system to compare daylight architectural spaces (Demers 2007). While this approach does not introduce empirical factors such as human preference, it does allow for the *relative comparison* of interior architectural spaces through methods such as average luminance and standard deviation. This research explored the *range* of daylight design strategies present within interior architecture and introduced a dialog about how we can contextualize and compare the visual effects of light (luminance). By extending the scope of research beyond tightly controlled side-lit office spaces, such as those studies presented in Sect. 2.2.3, we can begin to account for the complexity of visual effects that emerge from existing architecture.

2.3 Temporal Considerations for Daylight Performance

Section 2.2 introduced existing metrics for evaluating illumination and visual comfort for task-driven performance as well as research aimed at evaluating the perceptual field-of-view under daylight conditions. While the dynamics of daylight have influenced the development of annual climate-based illumination metrics such as daylight autonomy and visual comfort metrics such as annual daylight glare probability, there is a lack of consideration for temporal variability in those studies that evaluate the perceptual field-of-view.

Section 2.2.3 introduced existing methods for measuring luminance across our field-of-view, highlighting those methods that distinguish *spatial* diversity, such as the Luminance Differences index (Parpairi et al. 2002). However, we are still missing a method for measuring *temporal* diversity as it pertains to occupant satisfaction and human delight. Although HDR images can be used to quantify brightness and contrast in architectural space through luminance measurements, dynamic sky conditions necessitate a multitude of images, taken throughout the year, in order to account for the varied perceptual impacts of daylight through time.

One of the most challenging aspects of annual daylight analysis, whether it be luminance or illuminance based, is representing a large quantity of data simultaneously in both quantitative and visual terms. Spatio-Temporal Irradiation Maps (STIMAPS) were proposed as a way of representing annual data across a single graph, with days of the year on the horizontal axis and hours of the day on the vertical (Glaser and Hearst 1999) (Fig. 2.8). To help designers visualize the dynamic performance of daylight throughout the year, a simulation platform that combines ST maps with u-d goals and associated annual daylight renderings has been developed by Andersen and her research group, originally at MIT and now at EPFL (Andersen et al. 2013; Andersen, Gagne & Kleindienst, 2013; Kleindienst & Andersen, 2012, Gagne et al. 2011, Andersen et al. 2008).

This simulation method provides the designer with goal-based illuminance thresholds and allows them to navigate the resulting temporal maps alongside associated renderings. This provides a clear visualization of both the quality and quantity of light in a given space over time (Kleindienst et al. 2008; Lee et al. 2009) (Fig. 2.9). Although the 'smoothness' of any temporal map depends on the number of annual instances and the interpolation method between each data point, the method has been validated for illuminance across 56 annual periods representing 7 daily and 8 annual intervals (Kleindienst et al. 2008).

Although they have not yet been integrated, perceptual field-of-view metrics that rely on HDR images are well suited for the Lightsolve platform, which generates 56 annual images as parts of its goal-based analysis. To conduct an annual analysis of both *spatial* and *temporal* diversity in light across our field-of-view, it is important that any new metrics be represented through dynamic quantitative and visual means.

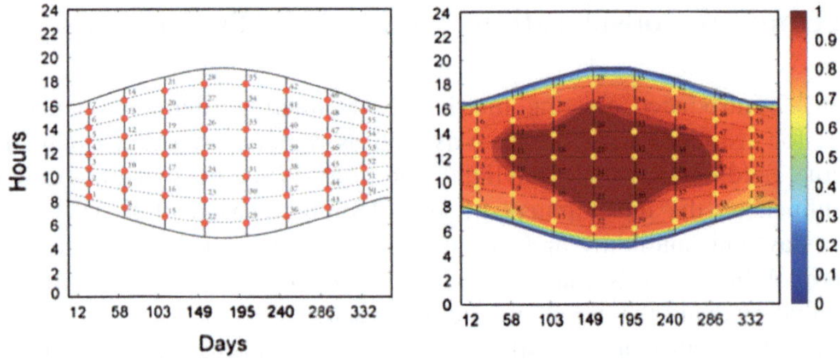

Fig. 2.8 Location of data points on a temporal map, 56 based on the temporal grid used in lightsolve

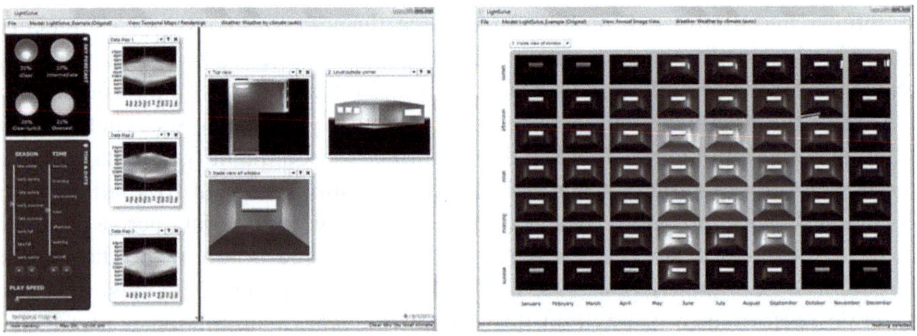

Fig. 2.9 Lightsolve interface, showing a default room with temporal illuminance maps on the top and annual renderings on the bottom (Kleindienst et al. 2008; Lee 2009)

2.4 Synthesis

Through a comparison of existing architectural spaces, this chapter introduced the importance of *spatial* and *temporal* diversity in our perception of daylight interior space. There are three categories that define existing daylight analysis metrics and methods: task-based illumination, visual comfort for task performance, and preferences toward the perceptual field-of-view. While task-based illumination metrics assess the amount of light required to perform visual task across a work plane, visual comfort metrics evaluate the potential for discomfort due to glare sources within an established view direction. Research directed toward the perceptual field-of-view has traditionally focused on brightness (mean luminance, threshold luminance, and luminance ranges) within a given view direction and occupant surveys to establish human preferences toward the luminous environment. Other studies of interest have coupled standard deviation (Wymelenberg and Inanici 2009) and/or visual *noise* (Parpairi et al. 2002) within an established view direction with occupant surveys to understand human preferences toward luminous diversity. While these studies begin to address the importance of *spatial diversity* in our perception of daylit space, they do not yet address the importance of *temporal diversity*, produced by the dynamics of sunlight throughout the year. The metrics proposed by this research will introduce a method for quantifying spatial contrast and luminous variability through the medium of digital images, so that these visual effects may be compared across a range of architectural spaces. It is the authors' perspective that existing task-based illumination and visual comfort metrics must be combined with dynamic perceptual metrics to create a more holistic understanding of daylight performance in architecture.

References

Andersen, M., Kleindienst, S., Yi, L., Lee, J., Bodart, M., Cutler, B. (2008). An intuitive daylighting performance analysis and optimization approach. Building Research and Information, vol. 36 (6), pp. 593–607

Andersen, M., Gagne, J.M.L., Kleindienst, S. (2013). Inter active expert support for early stage full-year daylighting design: a user's perspective on Lightsolve. Automation in Construction, in press.

Andersen, M., Guillemin, A., Amundadottie, M., & Rockcastle, S. (2013). Beyond illumination: An interactive simulation framework for non-visual and perceptual aspects of daylight performance. Chambery: IBPSA.

Cetegen, D., Veitch, J., & Newsham, G. (2008). View Size and Office Illuminance Effects on Employee Satisfaction. *Proceedings of Balkan Light*, (pp. 243–252). Ljubljana, Slovenia.

CIE. (1926). *Commision Internationale de l'Eclairage Proceedings, 1924.* Cambridge: Cambridge University Press.

Cuttle, C. (2010). Towards the third stage of the lighting profession. *Lighting Research & Technology, 42*, 73–93.

Demers, C. (2007). A classification of daylighting qualities based on contrast and brightness analysis. *Conference Proceedings of the American Solar Energy Society*, (pp. 243–252). Cleveland, Ohio.

Gagne, J.M.L., Andersen, M., Norford, L. (2011). An Interactive Expert System for Daylighting Design Exploration, Building and Environment, vol. 46 (11): pp. 2351–2364.

Glaser, D., & Hearst, M. (1999). Space series: Simultaneous display of spatial and temporal data. In *Conference Proceedings of the IEEE Symposium on Information Visualization.* San Francisco.

Holl, S. (1999). *The chapel of St. Ignatius*. New York: Princeton Architectural Press.

http://www.diva-for-rhino.com. (2009). Retrieved from DIVA-for-Rhino.

http://www.rhino3d.com. (2007). Retrieved 2010, from Rhinoceros.

IESNA. (2000). *IESNA lighting handbook: Reference and application*. New York: Illuminating Engineering Society of North America.

Jakubiec, J., & Reinhart, C. (2012). The 'adaptive zone'—A concept for assessing discomfort glare throughout daylit spaces. *Lighting Research and Technology, 44*, 149–170.

Kleindienst, S., & Andersen, M. (2012). Comprehensive annual daylight design through a goal-based approach. *Building Research & Information, 40*(2), 154–173.

Kleindienst, S., Bodart, M., & Andersen, M. (2008). Graphical representation of climate based daylight performance to support architectural design. *LEUKOS, 5*(1), 39–61.

Lam, W. (1977). *Perception and lighting as formgivers for architecture*. New York: McGraw Hill.

Lee, E., Clear, R., Ward, G., & Fernandez, L. (2007). Commissioning and verification procedures for the automated roller shade system at the New York Times Headquarters. http://windows.lbl.gov/comm_perf_nyt_pubs.html. New York.

Lee, J., Andersen, M., Sheng, Y., & Cutler, B. (2009). *Goal-based daylighting design using an interactive simulation method*. Glasgow: Building Simulation.

Loe, D., Mansfield, K., & Rowlands, E. (1994). Appearance of lit environment and its relevance in lighting design: Experimental study. *Lighting Research and Technology, 26*, 119–133.

Mardaljevic, J. (2000). Simulation of annual daylighting profiles for internal illuminance. *Lighting Research and Technology, 32*(3), 111–118.

Moon, P., & Spencer, D. (1942). Illumination for a nonuniform sky. *Illuminating Engineering, 37*(10), 707–726.

Nabil, A., & Mardaljevic, J. (2006). The useful daylight illuminance paradigm: A replacement for daylight factors. *Energy and Buildings, 38*(7), 905–913.

Parpairi, K., Baker, N., Steemers, K., & Compagnon, R. (2002). The luminance differences index: A new indicator of user preferences in daylit spaces. *Lighting Research and Technology, 34*(1), 53–68.

Reinhart, C., & Voss, C. (2003). Monitoring manual control of electric lighting and blinds. *Lighting Research and Technology, 35*(3), 243–250.

Reinhart, C., & Walkenhorst, O. (2001). Validation of dynamic radiance-based daylight simulations for a test office with external blinds. *Energy and Buildings, 33*(7), 683–697.

Reinhart, C., Mardaljevic, J., & Rogers, Z. (2006). Dynamic daylight performance metrics for sustainable building design. *Leukos, 3*(1), 1–25.

Rogers, Z. (2006). Daylighting metric development using daylight autonomy calculations in the sensor placement optimization tool. Boulder, Colorado: Architectural Energy Corporation, http://www.archenergy.com/SPOT/download.html.

Steane, M. A., & Steemers, K. (2004). *Environmental diversity in architecture*. New York: Spoon Press.

Tiller, D., & Veitch, J. (1995). Perceived room brightness: Pilot study on the effect of luminance distribution. *Lighting Research and Technology, 27*(2), 93–101.

Ursprung, P. (2002). *Herzog & De Meuron: Natural history*. Montreal: Canadian Centre for Architecture.

Veitch, J., & Newsham, G. (2000). Preferred luminous conditions in open plan offices: Research and practice recommendations. *Lighting Research and Technology, 32*, 199–212.

Waldram, P. (1950). *A measuring diagram for daylight illumination for the measurement, predetermination and representation of natural lighting*. London: Batsford.

Ward, G. (1994). The RADIANCE Lighting Simulation and Rendering System. In *Proceedings of '94 SIGGRAPH Conference*, (pp. 459–472).

Wienold, J. (2009). Dynamic daylight glare evaluation. In *Proceedings of International IBPSA Conference.*

Wienold, J., & Christoffersen, J. (2006). Evaluation methods and development of a new glare prediction model. *Energy and Buildings, 38*(7), 743–757.

Wymelenberg, K., & Inanici, M. (2009). A study of luminance distribution patterns and occupant preference in daylit offices. *PLEA2009—26th Conference on Passive and Low Energy Architecture.* Quebec City.

Chapter 3
Architectural Context

Keywords Daylight architecture · Architectural design · Design typologies · Architectural matrix · Contrast · Light variability

3.1 Developing a Typology for Daylight Architecture

The previous chapter began with a critical look at existing daylight performance metrics and strategies for evaluating brightness and contrast within architectural space. We then presented the need for more visually dynamic and spatially dependent methods for quantifying contrast and temporal variability in order to develop a more holistic set of daylight performance criteria. We will now turn to existing architectural examples to develop a more effective typological vocabulary about the role of contrast and temporal variability. Given the interdisciplinary nature of this research and its aim of transcending the boundaries between design and environmental analysis, we began with examples of architectural design and worked backwards toward a quantitative method of analysis. A global survey of existing architecture was conducted to establish a range of daylight design strategies; these varied from direct and variable to diffuse and uniform interior lighting schemes. This survey led us to the development of a linear classification strategy for the perceived degree of contrast and hypothesized temporal variability present within each space. These categories were then distilled down into a series of case study spaces and digitally modeled to create a set of annual renderings. The quantitative methods for evaluating contrast and temporal variability, which will be introduced in more depth in the following chapter, emerged out of a range of perspectives about the distinguishing characteristics of each space.

In order to understand the varied characteristics of contrast that occur within daylit space, a number of contemporary architectural examples were analyzed to produce a matrix of typological conditions. Each example was studied using the trained intuition of an architect and then positioned within a linear gradient to represent the degree of contrast within each space and the degree to which those

S. Rockcastle and M. Andersen, *Annual Dynamics of Daylight Variability and Contrast*, SpringerBriefs in Computer Science,
DOI: 10.1007/978-1-4471-5233-0_3, © The Author(s) 2013

levels were anticipated to change over time. The left side of this gradient was meant to contain highly variable and contrasted daylight strategies while the right side was reserved for minimally variable, low-contrast strategies. This typological approach was necessary to establish an eventual method for quantifying contrast, because it allowed us to understand the gradient of possible daylight strategies and to develop a numerical scale against which each space could be compared.

3.2 The Architectural Matrix

The first architectural examples illustrate clearly opposed contrast characteristics that establish a high and low for each end of the intuitive contrast spectrum. The first example to emerge on the far-left or 'high-contrast' side of the spectrum is Santiago Calatrava's Milwaukee Art Museum (Fig. 3.1). The atrium located beneath the central structural 'wings' allows for direct sunlight penetration through a highly articulated glass roof. This space represents a high degree of contrast and temporal variability as sunlight moves across the overhead structure, adjusting the pattern of incoming light onto the walls and floor. On the far-right or 'low-contrast' side of the spectrum, is the Modern Art Gallery in Renzo Piano's addition to the Chicago Art Institute (Fig. 3.2). The double-layered roof that covers this gallery consists of metal louvers that block direct sunlight and translucent glass that diffuses indirect light, while vertical fenestration is controlled through a series

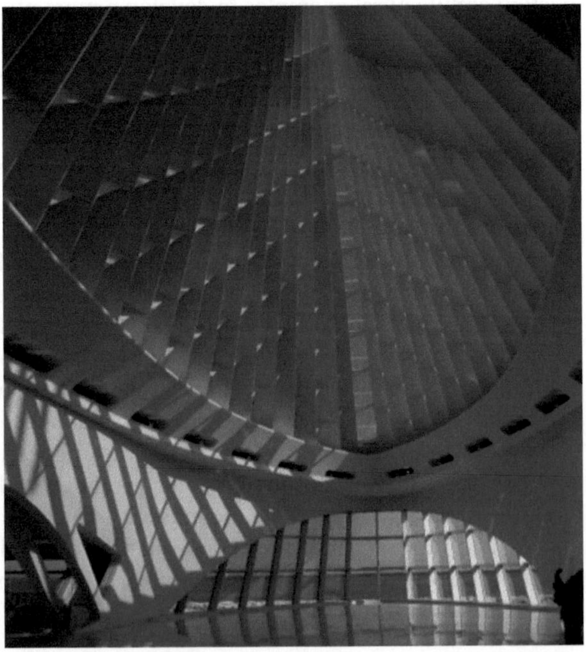

Fig. 3.1 Milwaukee Art Museum Kke227, 'Milwaukee Art Museum' October 28, 2007 via flickr, creative commons license

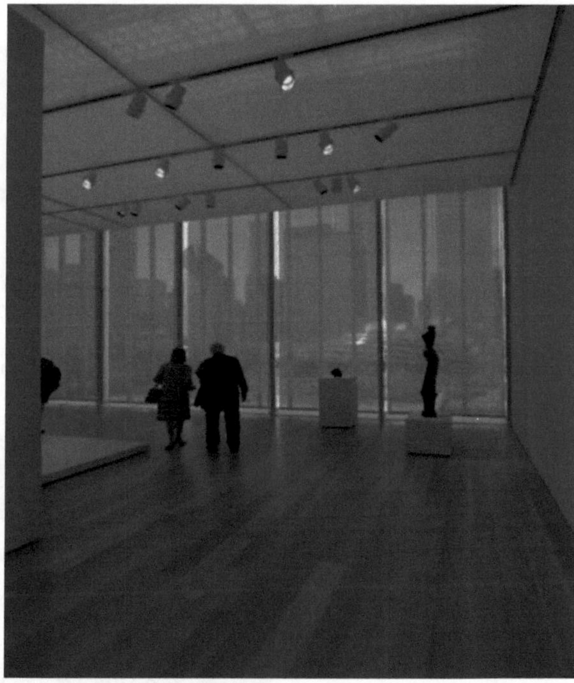

Fig. 3.2 Chicago Art Institute Ben B Miller, 'Southern View 3' May 12, 2009 via flickr, creative commons license

of roller shades. This space represents a low level of contrast as well as a low level of temporal variability due to the diffusing characteristics of its design strategy.

3.2.1 The Preliminary Matrices

The initial matrix positioned these two examples at each end of the spectrum and was composed of eight total categories that ranged from high contrast on the left to low contrast on the right (Fig. 3.3). The titles for these categories were, in this rendition, a work-in-progress, but they describe the qualitative differences between each column. At this time, we developed the term 'spatial contrast' to distinguish between various daylight characteristics; it is illustrated through a comparison of the Zollverein School of Management (Fig. 3.4), the Church of St. Ignatius (Fig. 3.5), and the Dia Beacon Museum (Fig. 3.6). All three spaces show some level of contrast between dark and bright areas within the image, although the Zollverein School of Management has sharper and more frequent spatial subdivisions or 'peaks' in the brightness between light and dark areas. The Church of St. Ignatius creates a more 'linear' or figural division between light and dark, whereas the Dia Beacon Museum has much smoother gradation and fewer spatial subdivisions.

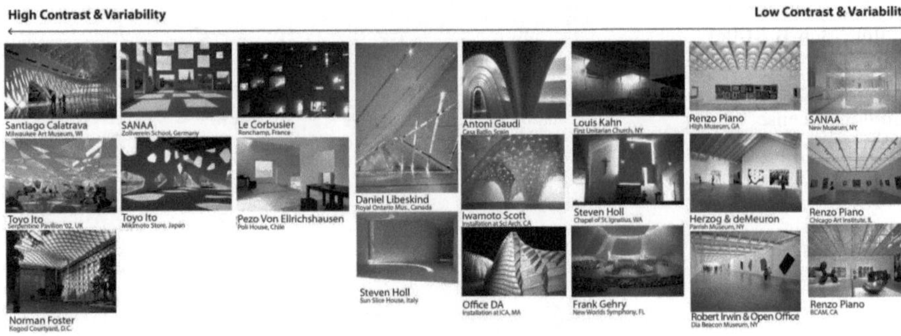

Fig. 3.3 Preliminary architectural matrix based on contrast and daylight variability

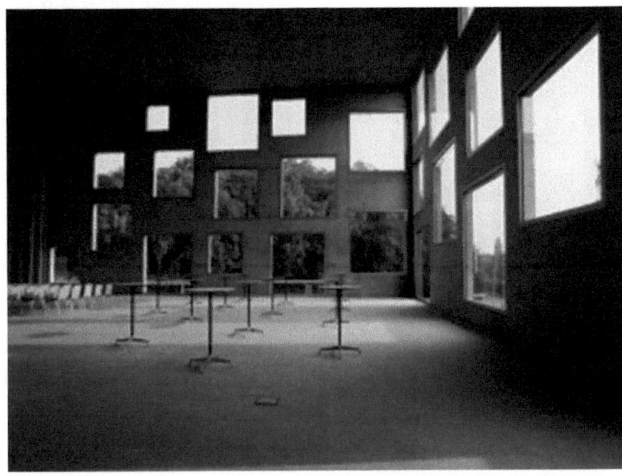

Fig. 3.4 Zollverein School Alena Hanzlova, 'Sanaa Zollverein School' October 12, 2007 via wikimedia commons, creative commons license

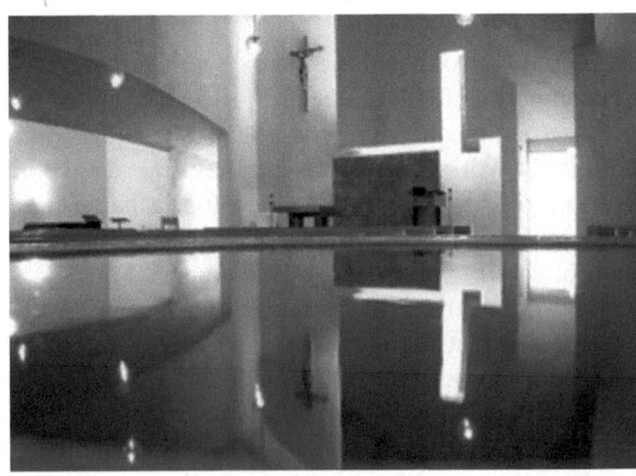

Fig. 3.5 Church of St. Ignatius, Joe Mabel, 'Chapel of St. Ignatius' November 30, 2007 via wikimedia, creative commons license

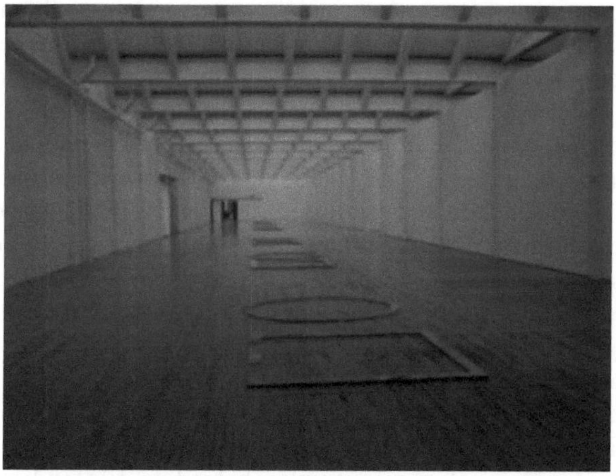

Fig. 3.6 Dia Beacon yusunkwon, 'Untitled' August 20, 2004 via flickr, creative commons license

This concept is illustrated in Fig. 3.7, which abstracts each space into a simplified model and each model into a map of enlarged pixels to represent the distinction between 'peaks,' 'lines,' and 'gradients.' Each cluster of contrast is conveyed by a field of circles that represent the strength in brightness of each pixel from 0 (black) to 255 (white). The thick circles represent pixel values closer to 255, while the thin circles represent values closer to 0. Spatial contrast is determined by the difference in brightness between neighboring pixels and can be seen by how sharply the values drop off, creating more abrupt figural breaks or smoother gradients. When a cluster of thick pixels is surrounded by a perimeter of thin circles, then 'peaks' of contrast are present. When a field of circles shows little variation in thickness, then it represents a smooth 'gradient' of contrast. The spaces that populate the left side of this initial matrix display sharper 'peaks' of contrast while those on the right side show smoother 'gradients' between areas of brightness. The spaces that occupy the middle set of categories represent some combination of the two, including more figural 'lines' or distinct and isolated shapes of light.

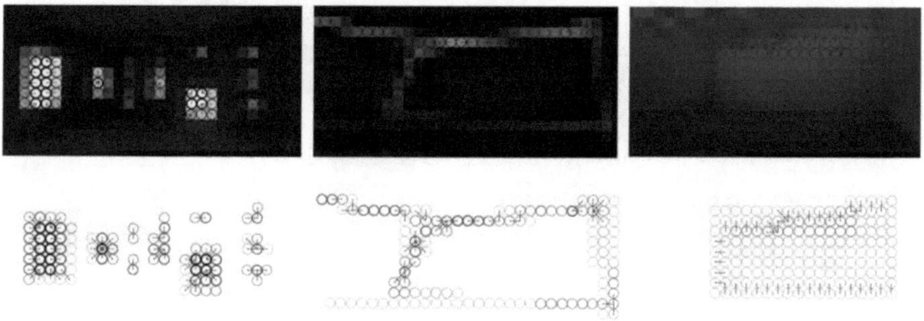

Fig. 3.7 'Peaks', 'lines,' and 'gradients' (from *left* to *right*)

This preliminary attempt at categorizing architectural space through its daylight characteristics relies on a certain amount of intuition from the perspective of a trained architect. The goal was not to place value on either side of the contrast spectrum, but to distinguish between contrast-driven effects to better understand how they might be defined more explicitly. The naked eye can analyze a photograph and identify the presence and location of contrast within space, but in order to understand the magnitude and stability of contrast as it changes over time, it is necessary to establish a framework through which spaces can be compared. This allows the designer a scale on which to locate and describe a desired effect, giving them a comprehensive understanding of contrast and its dynamic impacts. A typological approach to categorizing these effects brings us one step closer to quantifying the conditions they represent.

Due to the limited number of examples represented in the initial matrix, a reiterative strategy unpacked and expanded it to accommodate a broader set of categories. The addition of new examples helped to test and strengthen each category, raising the need for additional columns when new strategies emerged. This second matrix represents a more in-depth survey of architectural spaces, doubling the number of examples to forty-two and adjusting the total range of categories from eight to eleven (Fig. 3.8). This expanded matrix shifted examples from within its organization to develop a more resolved set of categories. Some spaces that were originally located on the left side of the spectrum were moved closer to the right as our notion of 'spatial contrast' began to distinguish between boundary conditions within the image. For example, the column containing the Poli House by Pezo Von Ellrichausen, originally located on the right side of the initial matrix, moved toward the center of the second matrix. This adjustment occurred when it was determined that the bright window openings did not create sharp peaks or hard boundaries of contrast against the interior space, as was seen in the Royal Ontario Museum. On the contrary, the thickness of the wall cavity in

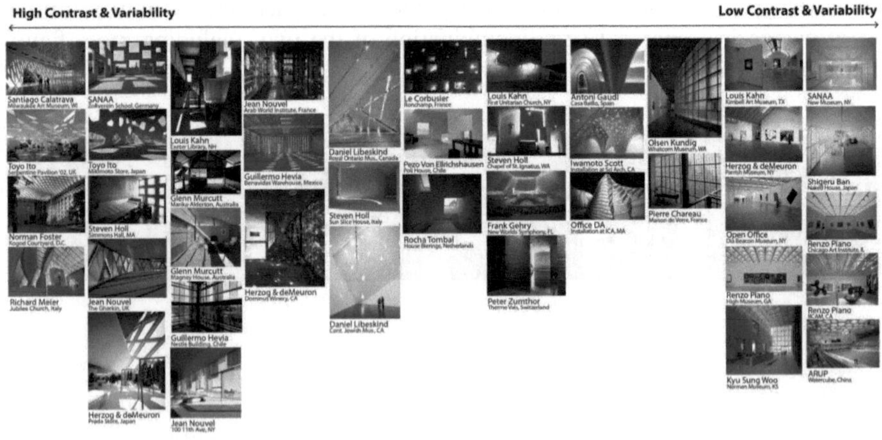

Fig. 3.8 Intermediate architectural matrix based on contrast and daylight variability

which the openings are set creates a smoother gradient of light as it enters the space. This can be seen by the tonal variations surrounding each window opening. The effect is a direct and indirect penetration of light, which is more similar in contrast to the First Unitarian Church, located in column seven, than it is to the Royal Ontario Museum, located in column two. This expanded matrix represents a process of trial and error that occurred throughout the development of this typological study. In order to define a set of qualitative and subjective principles, such as contrast, we must anticipate a certain level of resistance and a healthy degree of debate. This spirit of collaboration enables us to transcend the boundaries between architecture and technology to establish a new set of metrics that are dedicated to the values of both disciplines.

The most difficult spaces to define in this second matrix are located within the third column, which is titled Indirect and Selectively Direct. Upon further review, we determined that this category could be divided into two separate columns, each of which should be located toward the center of the matrix. Glen Murcutt's Magney House with its distinctive louvers and direct pattern of resulting sunlight represents more spatial contrast than others that were originally located within the same category. Jean Nouvel's 11th Avenue building employs a combined strategy of direct and indirect light penetration, similar to the Magney House, but it uses screens rather than louvers, which emit light in a softer set of gradients. These middle categories, many of which represent hybrid daylight strategies and combined contrast effects, are often difficult to distinguish and even harder to define. While it is impossible to categorize all examples of architecture through such explicit terms, the intention of this research is to generate a gradient of typological conditions against which similar characteristics can be compared. An overall range of contrast is established, despite some flexibility between adjacent categories. This typological comparison, however, subjective through the development of each category, establishes an original attitude toward the use of contrast and temporal diversity in evaluating daylight in architecture. This approach may be used to establish more objective criteria for the analysis of environmental performance in architecture as design intentions must be taken into consideration before metrics can be applied for evaluation.

3.2.2 The Full Matrix

Using the first two matrices as grounds for discussion and refinement, we created the third and final matrix of existing architectural spaces. The matrix contains seventy-five examples and spans fifteen categories, creating a more articulated gradient of contrast-driven effects (Fig. 3.9). An increase in the overall number of examples allows for more accurate differentiation between columns, although there are an uneven number of examples in each category as some typologies are more common than others.

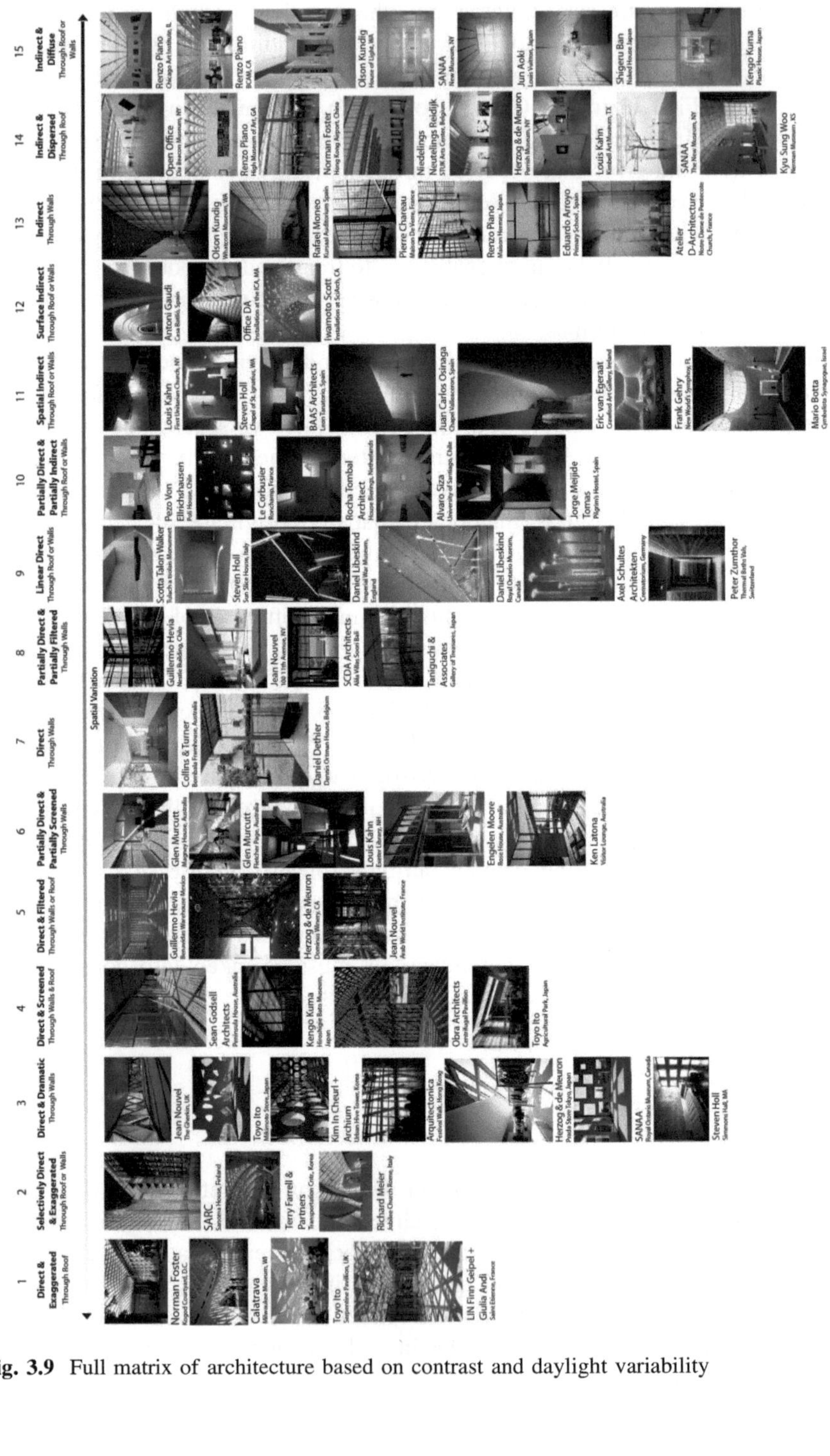

Fig. 3.9 Full matrix of architecture based on contrast and daylight variability

The first three categories, referred to (from left to right) as '*Direct and Exaggerated*,' '*Selectively Direct and Exaggerated*,' and '*Direct and Dramatic*' represent the far-left or high-contrast end of the spectrum. The '*Direct and Exaggerated*' column contains those spaces with transparent, top-lit daylighting strategies in which the '*direct*' penetration of sunlight plays a dominant role in the choreography of visual effects. It includes the Smithsonian Courtyard by Norman Foster and the Serpentine Pavilion by Toyo Ito. The next column, referred to as '*Selectively Direct and Exaggerated*,' describes similar contrast characteristics, but accommodates those spaces that have some opacity in their structure as can be seen in the Millennium Church by Richard Meier. The third column, known as '*Direct and Dramatic*,' includes spaces like the Prada Store by Herzog and De Meuron and the Zollverein School of Management by SANAA. The architecture in this category is defined by direct sunlight through an articulated transparent façade and displays high spatial contrast. The obvious differences between the first and third column are due to the orientation of the transparent light-emitting surface. When unobstructed sunlight is allowed in through the roof, it creates contrast on all four walls as well as on the floor. When it enters through the wall, it can only affect three vertical surfaces and the floor, reducing the overall contrast perceived within the space.

The next three categories, (from left to right) '*Direct and Screened*,' '*Direct and Filtered*,' and '*Partially Direct and Partially Screened*,' represent the high-to-middle portion of the contrast spectrum. The fourth column, '*Direct and Screened*,' contains the Centrifugal Pavilion by Obra Architects which represents those spaces with smaller gauge surface openings, resulting in some direct and some indirect light. The fifth column, '*Direct and Filtered*,' is similar in definition to the previous column, except that it contains spaces such as the Dominus Winery by Herzog and De Meuron which is defined by a smaller and less frequent pattern of incoming light. The sixth column, '*Partially Direct and Partially Screened*,' is characterized by the presence of a fully glazed façade which filters light through a set of louvers such as the Magney House by Glen Murcutt.

The middle three categories are labeled (from left to right) '*Direct*,' '*Partially Direct and Partially Filtered*,' and '*Linear Direct*.' While the column labels are self-explanatory, they are difficult to populate as the spaces that fall within them represent some form of hybrid contrast. The seventh column, '*Direct*,' includes spaces such as the fully glazed Bombala Farmhouse by Collins and Turner. This category is defined by fully glazed, side-lit conditions that allow for maximum sun exposure with minimal obstruction. Category eight is distinguished by the presence of a fully glazed façade which filters light through a smaller gauge screen such as the Nestle Social Building by Guillermo Hevia Architects. The last category in this group, '*Linear Direct*,' is composed of spaces like Daniel Libeskind's Imperial War Museum, which emits light through clearly defined slit(s) in the wall which result in dramatic and figural shapes.

The middle-to-low-contrast categories, '*Partially Direct and Partially Indirect*,' '*Spatial Indirect*,' and '*Surface Indirect*,' are composed of spaces that primarily emit indirect light with smoother gradients of contrast. The '*Partially Direct and Partially Indirect*' category includes architecture like the Poli House by Pezo Von Ellrichausen, which allows for the direct penetration of sunlight through thick outer wall openings. Instead of creating sharp contrast boundaries, these windows diffuse light through their depth. The next category, '*Spatial Indirect,*' operates in a similar way, but only through indirect light such as can be seen in the First Unitarian Church by Louis Kahn. The twelfth column, '*Surface Indirect,*' is made up of non-planar surface conditions that create indirect lighting patterns, providing low-level contrast. This column contains more computationally complex spaces such as the Sci-Arch installation by Iwamoto Scott Architects.

The final set of categories in this matrix, labeled '*Indirect,*' '*Indirect and Dispersed,*' and '*Indirect and Diffuse,*' represents the far-right or low-contrast end of the spectrum. All three columns are defined by indirect lighting strategies, from the Whatcom Museum by Olson Kundig to the Chicago Art Institute by Renzo Piano. The difference among these three columns is characterized by whether the space is side-lit or top-lit and by the degree of resulting surface articulation through mullions and other structures. The lowest contrast spaces, located within the '*Indirect and Diffuse*' column fifteen, are represented by translucent overhead lighting and minimal surface noise.

Although there was no initial bias in the specific programmatic use of each space and its location within the matrix, there are definite patterns that emerge as a result of this system of classification. Those spaces located to the far left of the matrix, falling under '*Direct and Exaggerated*' or '*Direct and Dramatic,*' tended to represent circulatory, atrium, or unspecified public uses. Those spaces located to the far right of the matrix, '*Indirect and Dispersed*' and '*Indirect and Diffuse,*' were almost all gallery spaces with highly specific lighting needs. Spaces that fell in the middle of the matrix under '*Selectively Direct*' or '*Partially Direct and Partially Screened*' represented a mixture of programmatic uses but were dominated by residential examples. Interestingly enough, many of the religious programmatic spaces and concert or performance venues fell in the second half of the matrix under '*Partially Direct and Partially Indirect*' and '*Spatial Indirect.*' It may be no surprise that certain trends emerged through this typological approach to categorizing contrast as there are intuitive rules of thumb for the appropriate use of direct and diffuse lighting strategies for various programmatic uses. For example, it would be inappropriate for a museum to employ a '*Direct and Exaggerated*' daylighting approach in its galleries as it would create figural conditions of light that would distract from the artwork. Likewise, there is no need to minimize incoming light through an atrium space that is often meant to provide a transitional and variable experience for its occupants, who may spend the rest of their day in an artificially controlled office environment. In either case, there are definite correlations between programmatic use and the use of contrast in this gradient of contrast-driven daylight strategies.

3.3 The Typological Matrix

The final matrix of contrast-driven architectural effects (Fig. 3.9) made it possible to distill each column down into a single representative space so that we might understand the gradient on a more explicit level. Each category of the matrix was compressed into a single model that represents the characteristics of the existing examples, but with abstracted levels of detail.

The examples represented in Fig. 3.10 were organized into fifteen categories based on the presence of redundant contrast characteristics. The resulting typological spaces were a simple reduction in those characteristics into abstracted digital models that share similar floor plan (30′ × 30′) and ceiling height (10′) dimensions for the ease of comparison. Our intention was to loosely model each typological space after an existing architectural space, as represented by the previous matrix. The digital models were built using Rhinoceros, a NURBS-based geometric modeling program that is used by students and professionals alike to generate a wide array of architectural forms (http://www.rhino3d.com, 2007; http://www.diva-for-rhino.com, 2009). In this initial set of typological models, little time was spent on recreating the complexity of materials or interior finishes that might contribute to the overall contrast characteristics.

The first three categories represent the high-contrast, high-variability end of the spectrum. Category one, '*Direct and Exaggerated*,' has a transparent glass roof structure and south-facing façade with a simple grid of mullions that cast harsh shadows onto the walls and floor. Category two, '*Selectively Direct and*

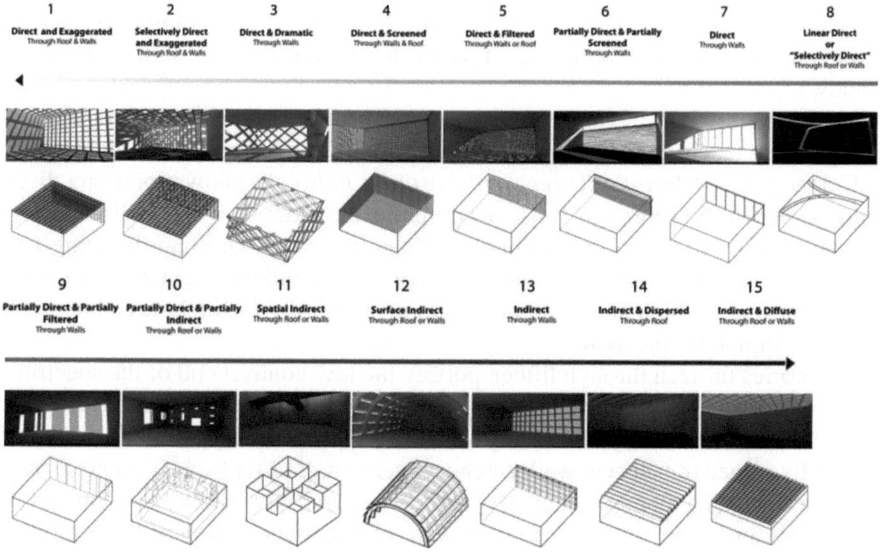

Fig. 3.10 The typological matrix

Exaggerated,' uses the same gridded structure as the previous space, but introduces some translucent glass panels to soften the incoming shadows. Category three, '*Direct and Dramatic*,' has a diagrid structure on all four walls to allow direct sun penetration in onto the floor. The difference between these categories comes from the location of openings in the roof and walls. Categories one and two are top-lit, which means that contrast and temporal variability will occur throughout the year, while category three is dependent on the time of day and year as sunlight must penetrate through vertical openings in the façade.

Category four, '*Direct and Screened,*' allows light in through both the walls and ceiling, while category five, '*Direct and Filtered*,' and category six, '*Partially Direct and Partially Screened*,' allow for varying degrees of direct penetration through a single vertical surface. Categories four and five have similarly scaled light-emitting openings, but category four casts an even array of light through holes in the roof and walls, while category five emits smaller, more sporadic pattern of light through a single wall. Category six is modeled to represent a clear-story window above with evenly space louvers below, emitting shadow lines onto the walls and floor. Due to the location of openings in category four, contrast is high throughout the year, while it varies with the seasons in category six, which determines its placement to the left of the group.

Categories seven through nine sit in the middle of the matrix and represent a moderate amount of contrast and temporal variation. Category seven, '*Direct,*' has one fully glazed wall that allows direct sunlight into the space, but does not experience the same degree of spatial contrast as those to the left due to its lack of shadow texture. Category eight, '*Selectively Direct,*' has isolated linear openings in the roof and walls, which allow for direct light penetration through isolated moments. This causes a high degree of spatial contrast, but only across a small percentage of the image. Category nine, has a fully glazed wall, similar to category seven, but combines some translucent panels to minimize the strength of incoming shadows.

Categories ten, eleven, and twelve move further toward the low-contrast end of the spectrum, each representing some technique for emitting indirect light.

Category ten, '*Partially Direct and Partially Indirect,*' allows for some direct sunlight penetration through thick openings in the exterior wall, while category eleven, '*Spatial Indirect,*' brings indirect light in through large north-facing roof monitors. Category twelve, '*Surface Indirect,*' emits indirect light onto a curved exterior surface, which shows some temporal variability in brightness, but stable contrast throughout the year.

Categories thirteen through fifteen portray the low-contrast end of the spectrum, with stable luminosity across the visual field. The difference between these can be seen in the location of light-diffusing surfaces, where category thirteen, '*Indirect*,' has a fully glazed translucent wall, category fourteen, '*Indirect and Dispersed,*' uses small, north-facing monitors, and category fifteen, '*Indirect and Diffuse,*' emits light through a translucent roof. Due to the nature of materials in each of these models, little temporal variation is expected throughout the year, and the visual field remains relatively stable.

The purpose of these typological models was to render each condition through a quick digital model so that more explicit readings could be determined from its abstract form. The next chapter will use this typological matrix to propose three new metrics that describe the elements of spatial contrast and temporal variability that have been explored through this gradient of contrast-driven daylight effects.

In this chapter, we have taken the need for new contrast-based metrics and positioned it within the context of contemporary architecture to develop a more comprehensive understanding of possible daylight-driven effects. A new typological language was developed to further understand this range of conditions; and each category was populated with existing examples to create a framework through which specific characteristics may be defined. The reduction in each category into an abstract typological model provides a spring board for the next chapter, which will propose three new and distinct metrics for the quantification and comparison of contrast and temporal variability in daylight architecture.

References

http://www.rhino3d.com. (2007). Retrieved 2010, from Rhinoceros.
http://www.diva-for-rhino.com. (2009). Retrieved from DIVA-for-Rhino.

Chapter 4
Defining New Metrics for Contrast and Variability

Keywords Spatial contrast · Annual spatial contrast · Annual luminance variability · Daylight architecture · Perceptual performance

In the previous chapter, we conducted a survey of contemporary global architecture to develop a typological language about the range of contrast-driven effects in daylight design. Each example was placed within a larger matrix to establish a gradient from high contrast on the left to low contrast on the right (Fig. 4.1). Using this system of categorization, we distilled each column of the matrix into a single abstract digital model to facilitate a more directed discussion about the specific characteristics that distinguish each space. We introduced new terms: 'spatial contrast' and 'temporal variability,' which describe intuitive differences in the distribution of daylight between each space and the overall placement of each example within the gradient from left to right. This chapter will examine those characteristics presented by the typological matrix and will introduce three new metrics: spatial contrast, annual spatial contrast, and annual luminance variability that contribute to a more holistic understanding of contrast and temporal variability in architecture.

4.1 Learning from the Typological Matrix

Daylight is a dynamic and highly variable natural source of illumination within architectural space. It reveals a variety of effects over time and cannot be understood through a singular instance or unit of measurement. The metrics introduced within this chapter will propose a new method for quantifying spatial contrast and temporal variability as well a framework for representing and comparing these variables across a range of architectural examples. In this section, we will discuss the placement of each architectural example within the typological gradient, introduced in Chap. 3. Our analysis will determine which luminous characteristics contributed to our initial categorization of each space. These

S. Rockcastle and M. Andersen, *Annual Dynamics of Daylight Variability and Contrast*, SpringerBriefs in Computer Science,
DOI: 10.1007/978-1-4471-5233-0_4, © The Author(s) 2013

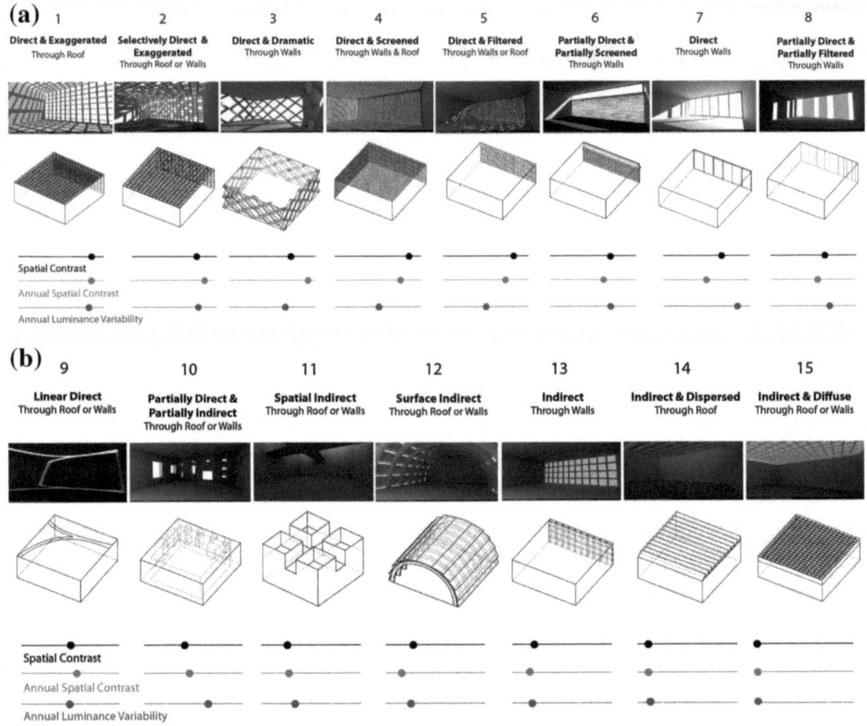

Fig. 4.1 Typological matrix with indicate intuitive scales for contrast and variability

characteristics will then be broken down into a quantifiable method for both spatial contrast and temporal variability.

To determine how qualitative effects regarding contrast and/or variability can be differentiated, we must first understand these concepts on an intuitive level. The typological matrix, re-introduced in Fig. 4.1, includes a set of sliding bars beneath each category. These sliders show the amount of *spatial contrast*, *annual spatial contrast,* and *annual luminance variability* hypothesized within each image. As intuitive concepts, these can be described as follows: *Spatial contrast* accounts for the strength of peaks, lines, and gradients within an image, while *annual spatial contrast* describes an accumulation of those values across the year. *Annual luminance variability*, on the other hand, describes the cumulative *variation* in brightness across the year. As a result, the position of each slider varies across categories and often displays different rankings for each of the three concepts. This approach was meant to separate the contributing characteristics of contrast and variability to understand them as a group of factors that work together in the production of daylight-driven effects.

The distinctions among these three characteristics are important to understand as each example within the typological matrix represents unique combinations of spatial contrast, annual spatial contrast, and annual luminance variability. In category ten, *partially direct and partially indirect*, for example, a fair amount of indirect light is emitted through each opening in the thick exterior wall, but receives a minimal amount of direct sunlight across the year and thus a low score for annual spatial contrast (Fig. 4.1). The space does, however, experience a fair degree of annual luminance variability as fluctuating light levels cause tonal variations on the walls and floor over time. This distinction is necessary when you consider spaces like Louis Kahn's First Unitarian Church, which allows for the penetration of indirect light through large, translucent roof monitors (Fig. 4.2, represented by category eleven, *spatial indirect*). There is never a high degree of spatial contrast present within the church, as daylight washes the walls in smooth gradients of illumination, but luminance levels still experience a high degree of variation, with fluctuating light conditions affecting the brightness of those gradients across the year.

Those spaces that receive a large quantity of direct sunlight generally result in high values for spatial contrast. It is important, however, to understand how those values vary across the day and year. A side-lit space, such as those described by categories six and seven, can achieve high spatial contrast during the winter months, when the solar altitude is lower to the ground, but may achieve a minimal degree of spatial contrast during the summer months. The dynamic nature of sunlight makes it critical to distinguish between static and annual representations of space. Architectural spaces such as the Denver Art Museum (Fig. 4.3, represented by category eight, *linear direct*) and the Zollverein School (Fig. 4.4, represented by category three, *direct and dramatic*) may experience large jumps in spatial contrast over time, as more direct sunlight is driven into the space depending on the orientation of light-emitting surfaces. Annual spatial contrast is useful in distinguishing between spaces that achieve high levels of contrast across the year and those that achieve it only intermittently.

Annual luminance variability describes the cumulative variation in brightness within architectural space as it varies from one moment to the next. Depending on the orientation of incoming light, this annual *climate-driven* metric can represent high degrees of variability even when spatial contrast levels remain low. In order

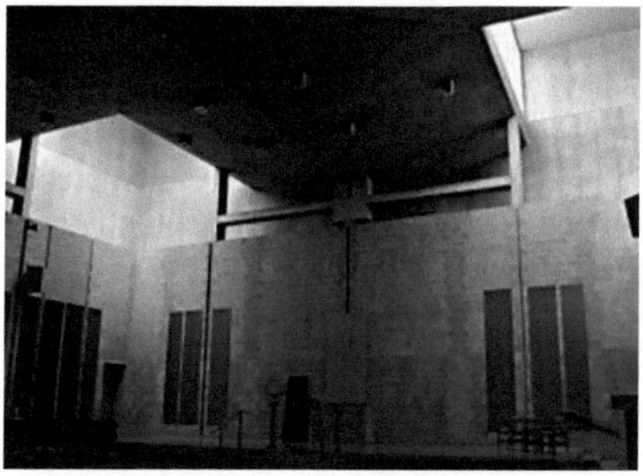

Fig. 4.2 First Unitarian Church © Bryan Maddock

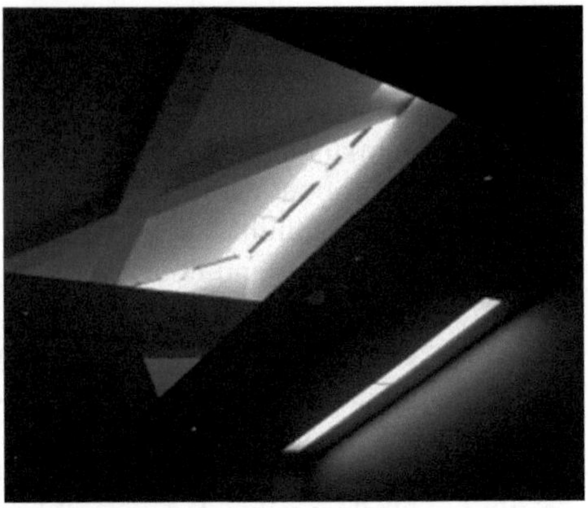

Fig. 4.3 Denver Art Museum, chad_k, September 1, 2007, via Flickr, Creative Commons License

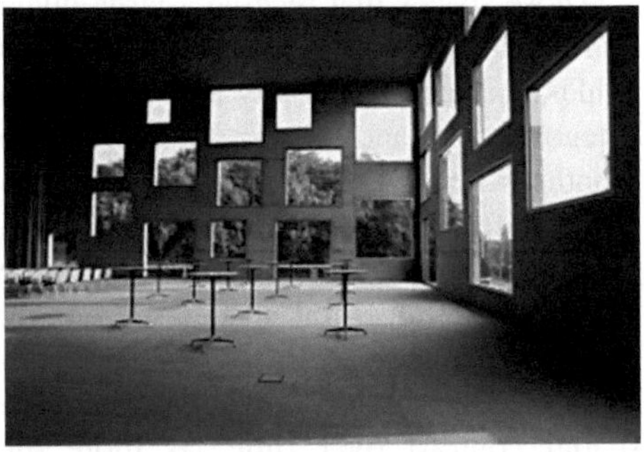

Fig. 4.4 Zollverein School Alena Hanzlova, 'Sanaa Zollverein School' October 12, 2007, via Wikimedia Commons, Creative Commons License

to account for the nuanced variations that can occur within daylit spaces, it is important to reference each of these metrics in order to gain a better understanding of their combined performance criteria.

4.2 Contrast and Variability Metrics

This section will examine three proposed metrics: spatial contrast, annual spatial contrast, and annual luminance variability. The specific characteristics of daylight that each metric seeks to address will be presented as well as the quantitative approach used to calculate and represent them. Those metrics that rely on an annual set of renderings or photographs will be explained through applications in the following chapter.

4.2.1 Spatial Contrast

Unlike more traditional methods of contrast analysis that rely on brightness ratios and/or standard deviation, spatial contrast proposes a compositionally dependent method for quantifying local variations in brightness within architectural space, which are perceptually dependent on their local surroundings.

Figure 4.5 illustrates this notion through the simple representation of black and white pixels. When the composition is split down the middle, with half the pixels representing RGB 0 (black) and the other half representing RGB 255 (white), the histogram shows two columns of brightness values on either side of the spectrum (0 and 255). If we rearrange the composition to create more perimeter area between white and black pixels, the histogram remains unchanged. The red values to the right of the figure, representing spatial contrast, show the differences between neighboring white and black pixels. In this case, the change in composition affects the difference between neighboring values, increasing the spatial contrast. This method of quantification illustrates the impacts of spatial composition on our perception of contrast—where the patterns generated by sunlight make an impact on our perception of architectural space. Figure 4.6 reiterates this method through a simple representation of peaks and gradients that occur as a result of the difference between neighboring values. Building upon the simple representations of black and white pixels shown in Fig. 4.5 and the peaks and gradients illustrated in Fig. 4.6, we will now look at a more detailed example that calculates spatial contrast across a larger image. Figure 4.7 contains a pixelated image of daylight within space and represents the local differences between the brightness of each pixel and that of its neighbor. If we add up all the local differences, represented in red, we can compute a total sum of difference across the image. The problem with this number, as it exists currently, is that it is dependent on the pixel density of the original image and cannot be numerically compared to images of a different density. To get around this issue, it is necessary to represent the metric as a ratio between the total difference in local values and the maximum difference that the image could achieve as a result of its pixel density. This ratio, expressed in red at the bottom of Fig. 4.7, represents spatial contrast as the difference between local pixel values in the image on the left over the 'maximum' checkerboard of black and white values on the right.

In order to apply this operation to images that represent a higher resolution of pixel density, spatial contrast is computed in MATLAB 2011 by importing each image and converting it into a two-dimensional grayscale matrix. In its current state, the program for spatial contrast reads jpeg images of any pixel density, but is also capable of processing HDR formats.

To explain the computational workflow in a comprehensive manner, we will use Fig. 4.8. Once an image file is imported into the MATLAB environment, the data are converted into a matrix of RGB values (between 0 and 255) that represent the brightness of each pixel. From there, we extract two new matrices, representing the difference between each row (shown in red) and column (shown in blue).

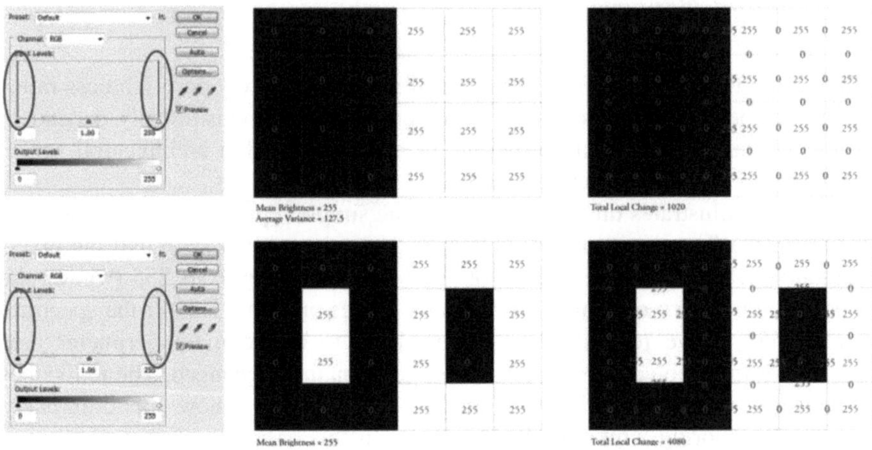

Fig. 4.5 Representation of RGB histograms and the impact of pixel composition on spatial contrast (seen in red values to the right)

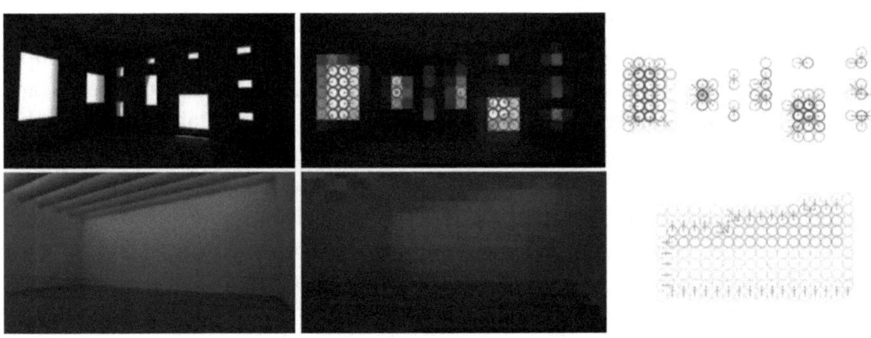

Fig. 4.6 Peaks and gradients of spatial contrast, represented by the strength of circles

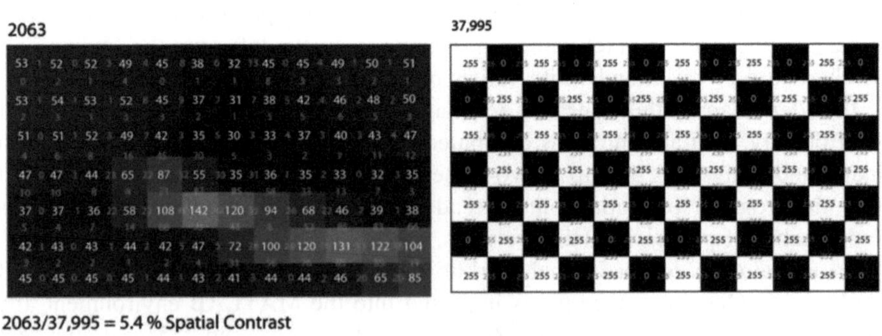

Fig. 4.7 Ratio of spatial contrast over a hypothetical high based on pixel density

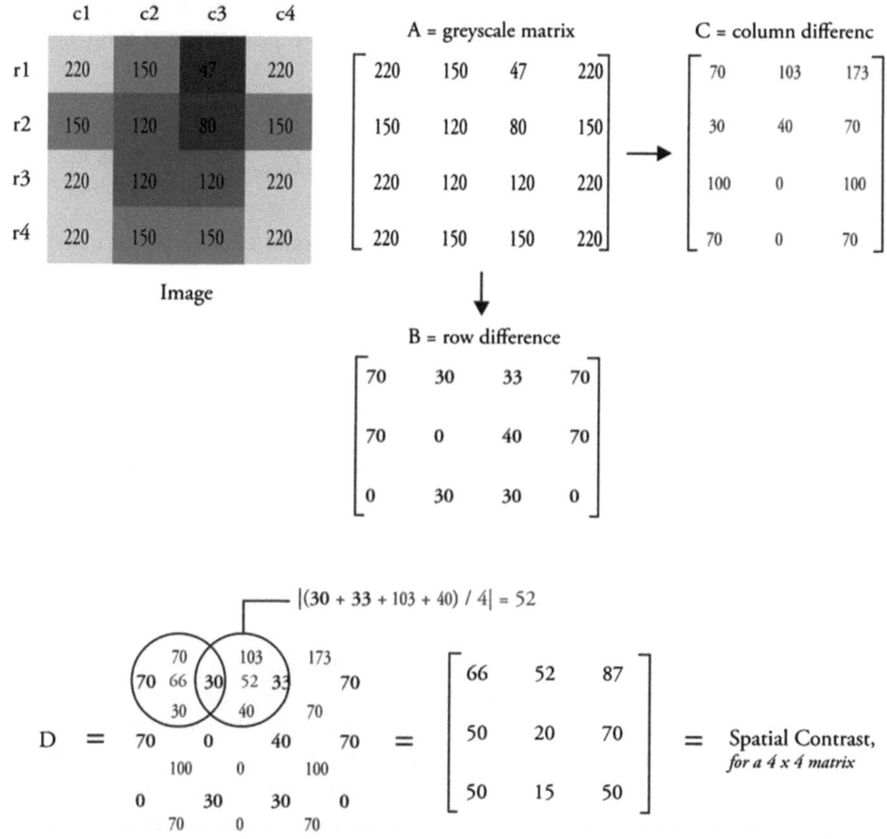

Fig. 4.8 MATLAB logic for spatial contrast

Although the conceptual diagram for calculating local differences seems straightforward, processing pixel information as a matrix requires us to deal with resulting matrices that D differ in size. For example, the original image matrix (shown in black) is 4 × 4 in dimension, whereas the resulting row difference matrix is 4 × 3 (shown in red) and the column difference matrix is 3 × 4 (shown in blue). The difficulty arises when we recombine the row difference and column difference matrices to create a data set that represents the original scale of the image. In order to achieve this, we chose to take the average of neighboring row and column differences (represented by the blue and red matrix) to create a rounded 3 × 3 matrix. This means that for an image with a pixel aspect ratio of 480 × 640, the resulting spatial contrast matrix will be 439 × 639 in size. The resulting ratio of spatial contrast can be computed using two matrices: one representing the original image and the other representing a black and white checkerboard of the same pixel density (see Fig. 4.7). Each matrix is processed in MATLAB to create an average of the row and column differences and then summed to produce a value for total spatial contrast and total hypothetical contrast. These two values are then turned into a ratio that represents the total spatial contrast of the image.

$$A_{n,m} \rightarrow B_{n-1,m}, C_{n,m-1} \quad E = \begin{bmatrix} 255 & 0 & \cdots & m \\ 0 & 255 & & \\ \vdots & & \ddots & \\ n & & & 255 \end{bmatrix} \rightarrow F_{n-1,m}, G_{n,m-1}$$

where A is a two-dimensional greyscale matrix and E is a two-dimensional checkerboard matrix of the same size that represents a hypothetical extreme…

$$b_{i,j} = \left|a_{i,j} - a_{i+1,j}\right| \qquad c_{i,j} = \left|a_{i,j} - a_{i,j+1}\right| \qquad \left\lceil d_{i,j} \right\rceil = \frac{\left(\frac{c_{i,j}+c_{i+1,j}}{2}\right) + \left(\frac{b_{i,j}+b_{i,j+1}}{2}\right)}{2}$$

$$f_{i,j} = \left|e_{i,j} - e_{i+1,j}\right| \qquad g_{i,j} = \left|e_{i,j} - e_{i,j+1}\right| \qquad \left\lceil h_{i,j} \right\rceil = \frac{\left(\frac{g_{i,j}+g_{i+1,j}}{2}\right) + \left(\frac{f_{i,j}+f_{i,j+1}}{2}\right)}{2}$$

$$\textbf{\textit{Spatial Contrast}}\ (\%) = \frac{\left(\sum_j \sum_i d_{i,j}\right)}{\left(\sum_j \sum_i h^f_{i,j}\right)} \times 100$$

Figure 4.9 illustrates these results: (a) shows a rendering on 28 November with a spatial contrast reading of 0.83 %, while (b) shows a rendering on 30 May with a spatial contrast reading of 0.97 %. In summary, the Spatial Contrast of an architectural rendering or photograph can be computed to produce a resulting black and white representation and a ratio, which can be used to compare it to the typological gradient of daylight effect introduced at the beginning of this chapter.

4.2.2 Annual Spatial Contrast

In order to understand the dynamic nature of sunlight and its changing impacts on architectural space, we created a second metric, annual spatial contrast, to quantify the *cumulative* effects of spatial contrast over time. Since daylight is a variable source of illumination, it is important that we develop metrics that can account for this variation on an annual scale.

In Chap. 2, we introduced a validated method for interpolating annual illuminance data across a set of 56 annual instances which correspond to 8 annual and 7 daily intervals (Kleindienst et al. 2008) This metric will analyze spatial contrast across these intervals, but rather than relying on weighted averages within each period according to weather conditions (Kleindienst et al. 2008), it will rely on clear sky conditions only and be evaluated for the corresponding instances (and associated sun positions). This variation can be considered as an upper boundary for contrast and variability as long as openings, depth, orientation, and positioning allow the

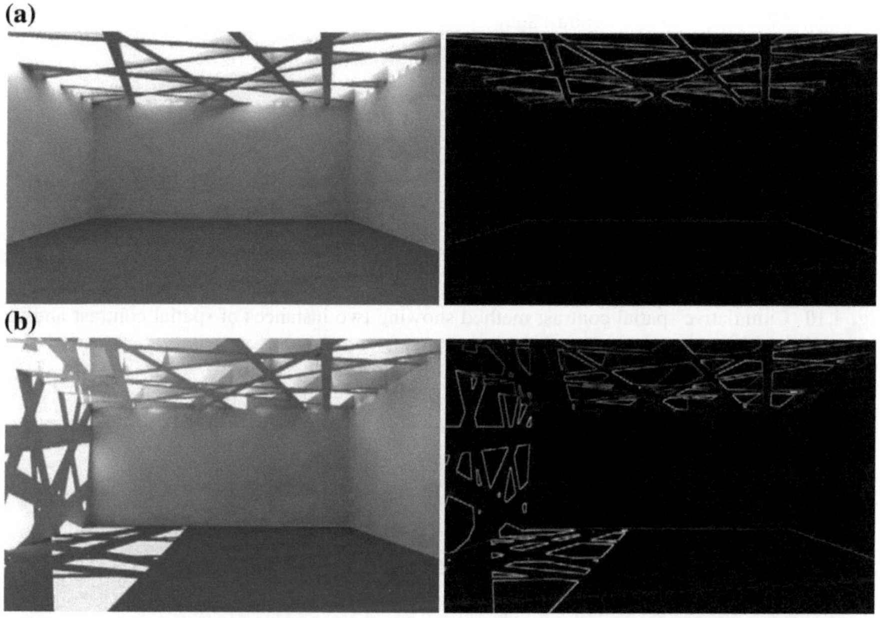

Fig. 4.9 **a** Rendering on 28 November showing a spatial contrast = 0.83 and **b** rendering on 30 May showing a spatial contrast = 0.97

limited set of 56 moments to reveal sun patches and adequate brightness. As such, it will provide a ratio for cumulative annual contrast as well as a spatio-temporal map of when that ratio changes over the year. The specific method for producing renderings across those 56 annual moments will be introduced in Sect. 5.2.

The basic process for calculating annual spatial contrast is shown in Fig. 4.10, which contains two pixelated RGB maps (one taken from a rendering at 10 am and the other at 10:30 am) with overlaid local contrast values in red. The sum of these values is added between each frame, representing a cumulative contrast sum across all 56 images. Figure 4.11 illustrates a full set of annual renderings (http://www.diva-for-rhino.com, 2009; http://www.rhino3d.com, 2007) (method of production described in Sect. 5.2) for a hypothetical space located in Boston, MA. To compute annual spatial contrast, this set of renderings is imported into MATLAB and processed by calculating spatial contrast for each of the 56 individual renderings and then layering that data accumulatively to visualize dynamic effects across the year.

Each of the individual spatial contrast ratios is plotted onto a spatio-temporal map for the latitude and longitude of Boston and represented alongside a false-color image that shows the location and intensity of cumulative contrast. Figure 4.12 illustrates three of these resulting spatial contrast representations (a, b, & c) and indicates their location on the spatio-temporal map below (seen in d). The false-color image on the lower right (e) shows the cumulative sum of all 56 spatial contrast

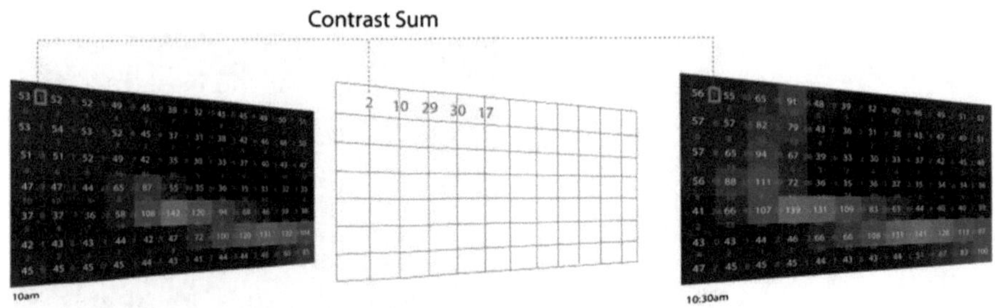

Fig. 4.10 Cumulative spatial contrast method showing two instances of spatial contrast and the sum of their values

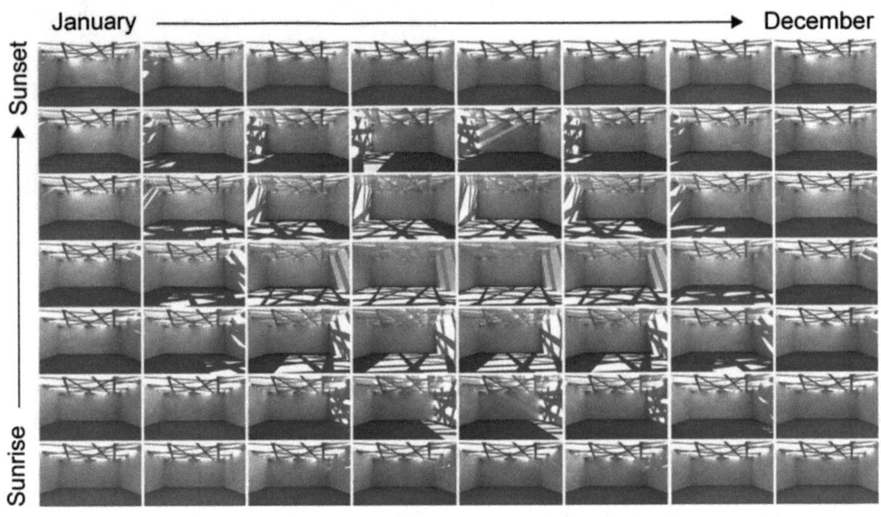

Fig. 4.11 56 annual images for Boston, MA (method of production explained in section Sect. 5.2), DIVA for Rhino, http://www.diva-for-rhino.com

images to illustrate the location and intensity of annual contrast within the space. It is most useful in examples in which the orientation of light-emitting surfaces creates dramatic seasonal variations, such as those spaces represented by categories one through eight in the typological matrix (Fig. 4.1). The numerical scale for spatial contrast has been determined by the results from a series of ten case studies, introduced in Chap. 5. Based on the distribution of values, two thresholds divide the data into three parts, each representing a third of the population (Fig. 5.6). As a result of this statistical subdivision, spatial contrast values between 0 and 0.5 are considered low, values between 0.5 and 0.8 are considered medium, and values exceeding 0.8 are considered high. To develop this metric through future research, a broader sample of existing architectural spaces will need to be modeled and analyzed to produce more statistically refined thresholds for spatial contrast. For now,

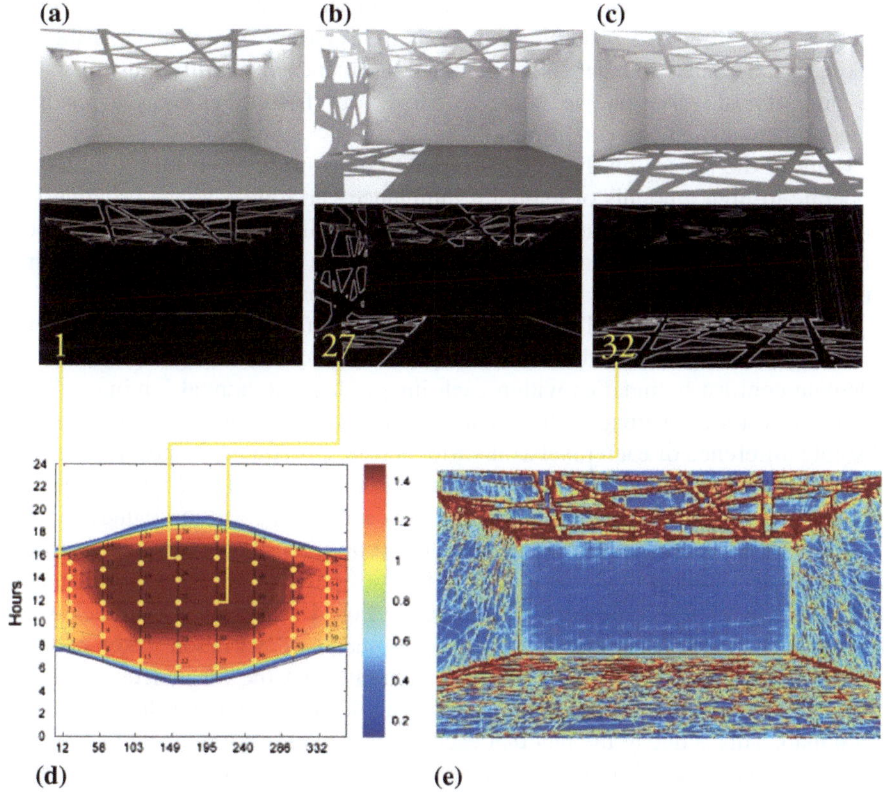

Fig. 4.12 Annual spatial contrast results for Case Study 1, *Direct & Exaggerated.* **a** rendering on 13 January at 10:33, **b** rendering on 30 May at 16:49, **c** rendering on 15 July at 12:50, **d** temporal map showing overlay of 56 plotted moments with the locations of a, b & c, and **e** cumulative image of annual spatial contrast

these thresholds represent a proof-of-concept and are used for relative comparison only between the case studies modeled in Chap. 5. The temporal map in Fig. 4.12 shows values that range from 0.6 (medium) to 1.4 (high).

4.2.3 Annual Luminance Variability

Annual luminance variability, which seeks to quantify the overall variation in brightness across an architectural space due to temporal fluctuations in daylight, is the third and final metric presented in this chapter. Whereas spatial contrast identifies compositional contrast boundaries within an image, and annual spatial contrast maps the accumulation of those contrast boundaries over time, luminance variability accounts for the accumulative differences in brightness *between* images.

Although one can measure instantaneous spatial contrast, luminance variability emerges as a difference in luminance levels *between* hourly and daily instances, making it inherently time dependent. Annual luminance variability thus represents the intensity of variation we perceive across our field-of-view as a result of dynamic annual lighting conditions.

Many of the spaces on the low side of the contrast spectrum may still display moderate-to-high amounts of temporal luminance variability, as the brightness of any given surface may transform dramatically while still maintaining smooth contrast gradients. This metric quantifies the degrees of light variation over time and allows us to see when and where those variations occur in space.

The quantitative method for this metric relies on the same set of 56 annual renderings that were introduced in the previous section; however, it does not calculate contrast boundaries within each image. Instead, annual luminance variability converts each image into a matrix of RGB values and then computes the absolute difference of each pixel as it varies across subsequent frames. Figure 4.13 illustrates how a single pixel can vary in brightness over time, while Fig. 4.14 shows how those differences can create a new matrix of resulting values.

Figure 4.15 exemplifies how the variation in brightness between four instances can produce a matrix of luminance variability through time. The four instances used to produce these values are represented in Fig. 4.16, which shows each of the 56 annual moments (a) and the resulting 42 moments of 'variation' taken between neighboring points (b). It is important to reiterate that the 56 images we use to calculate annual luminance variability result in 42 data points on the spatio-temporal map. This is due to the fact that each image represents a moment of time and this metric takes the difference *between* each of those moments. As a result, 7 daily intervals result in 6 daily points of total luminance variation and 8 seasonal intervals result in 7 seasonal points of total luminance variation (Fig. 4.16). The

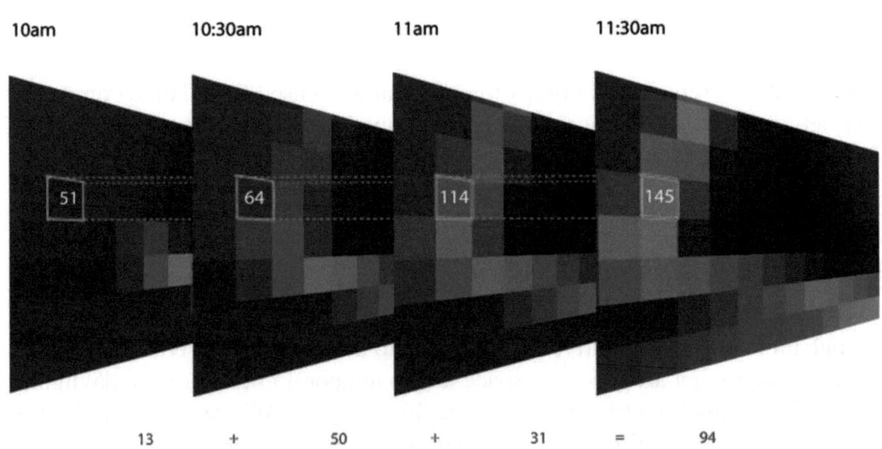

Fig. 4.13 Variability in the brightness of a single pixel over time (pixels are identified by a grayscale value between 0 and 255)

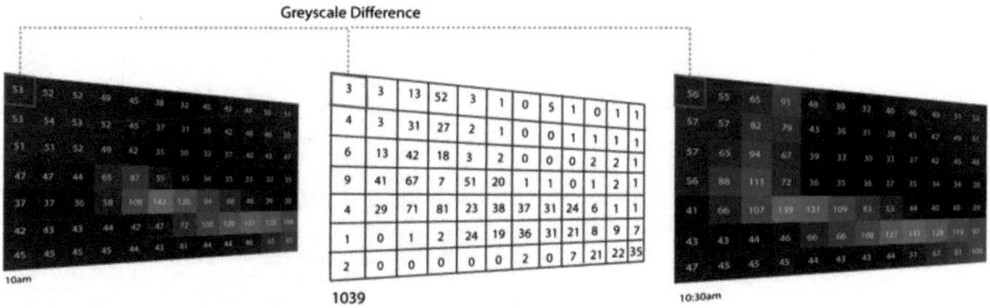

Fig. 4.14 Luminance variability method (pixels are identified by a value between 0 and 255)

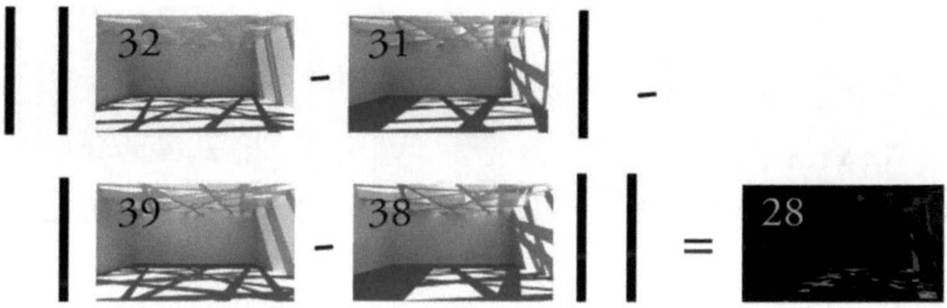

Fig. 4.15 Difference in luminance between four renderings. The date and time for each rendering is referenced in Fig. 4.16

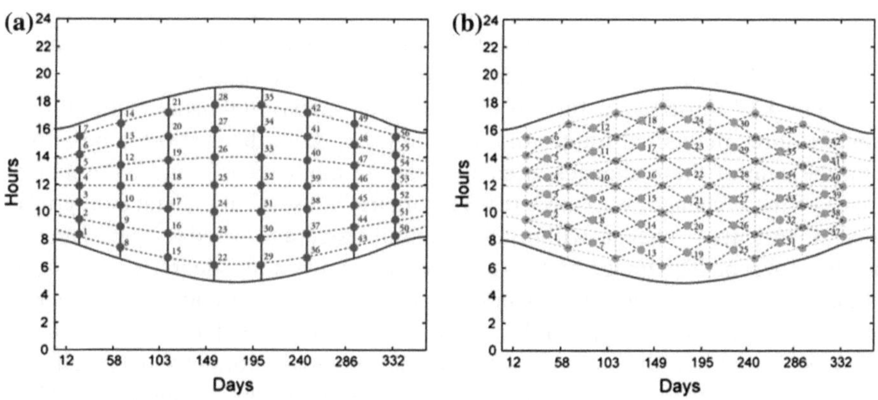

Fig. 4.16 Spatio-temporal map showing **a** the location of 56 data points for spatial contrast on the left and **b** 42 data points for annual luminance variability on the right

value for annual luminance variability is represented by the total sum of these 42 intervals. Similar to the spatial contrast, the resulting cumulative variation cannot be compared to images of varied pixel density until it is converted into a relative

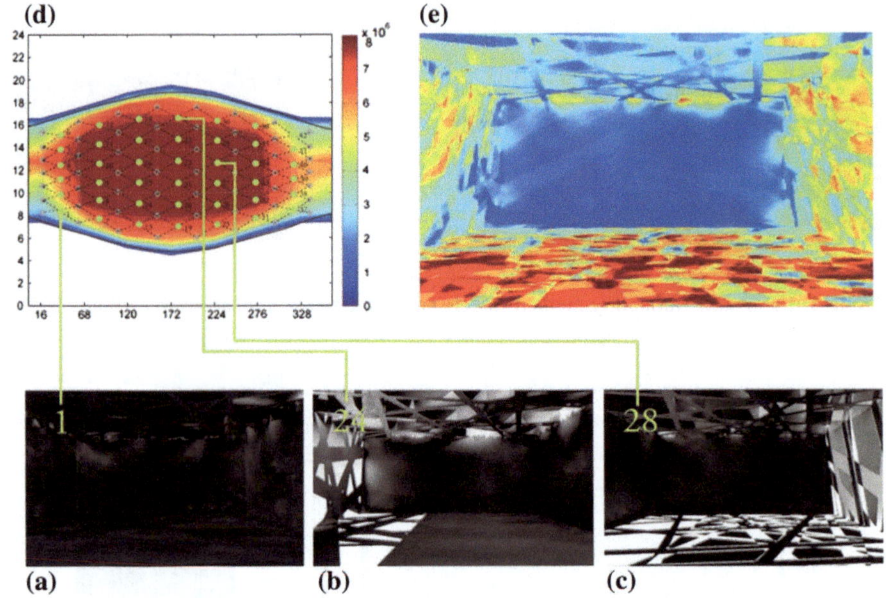

Fig. 4.17 Annual luminance variability. **a** Luminance variability for new data point 1 (taken between renderings 1, 2, 8, &9), **b** luminance variability for new data point 24 (taken between renderings 27, 28, 34, & 35), **c** luminance variability for new data point 28 (taken between renderings 32, 33, 39, & 40), **d** spatio-temporal map showing 42 point of luminance variability, and **e** accumulative image of annual luminance variability

value. In order to achieve this, the total sum of luminance variation across all 42 intervals is divided by the total pixel density of the input images.

Figure 4.17 illustrates a full set of results for annual luminance variability; it contains three individual frames of variation (a, b, & c), the spatio-temporal map with the sum of these changes at each of the 42 moments (d), and a cumulative image of these changes over time (e). This metric is useful in understanding *when* areas of brightness change within architectural space and whether that change is steady or abrupt. The image on the lower left (a) shows a low degree of luminance variability between renderings, while those to the right (b and c) show a high degree of variation. The temporal map (d) shows that these changes in luminance are most extreme in the summer when the sun is moving directly overhead. The cumulative image (e) shows where these variations occur within space. These changes appear to be most extreme on the floor, as direct light is constantly moving across the roof, casting variable patterns down into the room. Some change can also be seen on the walls, with minimal variation occurring in the roof, where light is always bright and minimally variable.

Like spatial contrast, the scale for luminance variability was established from the range of case studies introduced in Chap. 5. Based on the distribution of values, two thresholds divide the data into three parts, each representing a third of the population (Fig. 5.7) As a result of this statistical subdivision, luminance variability values between 0 and 2×10^6 are considered low, values between 2×10^6 and 3×10^6 are considered medium, and values exceeding 3×10^6 are considered high. To develop this numerical scale, future research will analyze a broader sample of existing architectural spaces to produce more statistically refined thresholds for luminance variability. Like spatial contrast, these thresholds represent a proof-of-concept and are used for relative comparison only between the case studies modeled in Chap. 5.

4.3 Synthesis

The previous chapter categorized contemporary architecture into a matrix of contrast-driven daylight strategies. This matrix was then distilled down into a set of abstract typological models in order to understand the characteristics that define each category and its location within the gradient. The present chapter has introduced three new metrics: spatial contrast, annual spatial contrast, and annual luminance variability which represent three distinct, yet related aspects of perceptual daylight performance and help to contribute to a more dynamic understanding of architectural space over time. In the following chapter, we will apply these metrics to a set of ten case study models and test their success in differentiating between those intuitive perceptual characteristics such as contrast and temporal variability that were discussed in Chapter 3. The results for these case studies will serve as a pre-validation for the proposed metrics.

References

http://www.diva-for-rhino.com. (2009). Récupéré sur DIVA-for-Rhino.

http://www.rhino3d.com. (2007). Consulté le 2010, sur Rhinoceros.

Kleindienst, S., Bodart, M., & Andersen, M. (2008). Graphical representation of climate based daylight performance to support architectural design. *LEUKOS, 5*(1), 39–61.

MATLAB. (2011). Récupéré sur http://www.mathworks.com/products/matlab/.

Chapter 5
Application of New Metrics to Abstract Spatial Models

Keywords Spatial contrast · Annual spatial contrast · Annual luminance variability · Case study analysis · Daylight design · Daylight simulation

In this Chapter, we will apply each metric introduced in Chap. 4 to a series of rendered case study spaces to test their success in quantifying and comparing spatial contrast, annual spatial contrast, and annual luminance variability. Although a total of fifteen categories were established in the typological matrix introduced in Chap. 3, we have selected ten case study spaces that represent an overall gradient of contrast and variability (Fig. 5.1). From these ten case study spaces, we have selected four examples to represent a cross-section of results in more detail, while the full spectrum of results for all ten case studies is presented more broadly toward the end of the Chapter.

5.1 Production of Annual Image Sets

In order to analyze annual spatial contrast and annual luminance variability within each case study, we must first produce a set of annual renderings or photographs that provide an accurate representation of the architectural qualities and daylight characteristics within each space. Figure 5.2 represents a workflow diagram showing three possible methods for the production of annual image sets.

The first method, which relies on Lightsolve, daylight analysis software developed first at MIT and more recently at EPFL, produces 56 annual renderings as part of its goal-based analysis (Kleindienst et al. 2008). These renderings are taken at symmetrical monthly and daily intervals for sunny, overcast, and turbid sky conditions to allow for an annual statistical climate-based analysis of Illuminance and DGP. The logic of Lightsolve's time segmentation method allows for a statistically representative sample of annual data while minimizing computational intensity.

S. Rockcastle and M. Andersen, *Annual Dynamics of Daylight Variability and Contrast*, SpringerBriefs in Computer Science,
DOI: 10.1007/978-1-4471-5233-0_5,

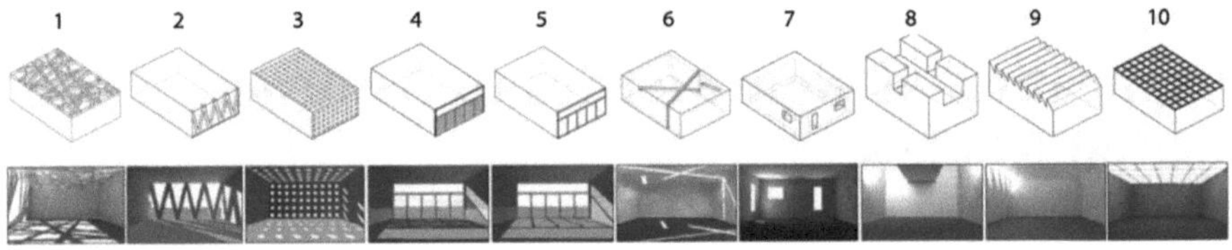

Fig. 5.1 Ten case study spaces, digitally modeled and rendered for analysis

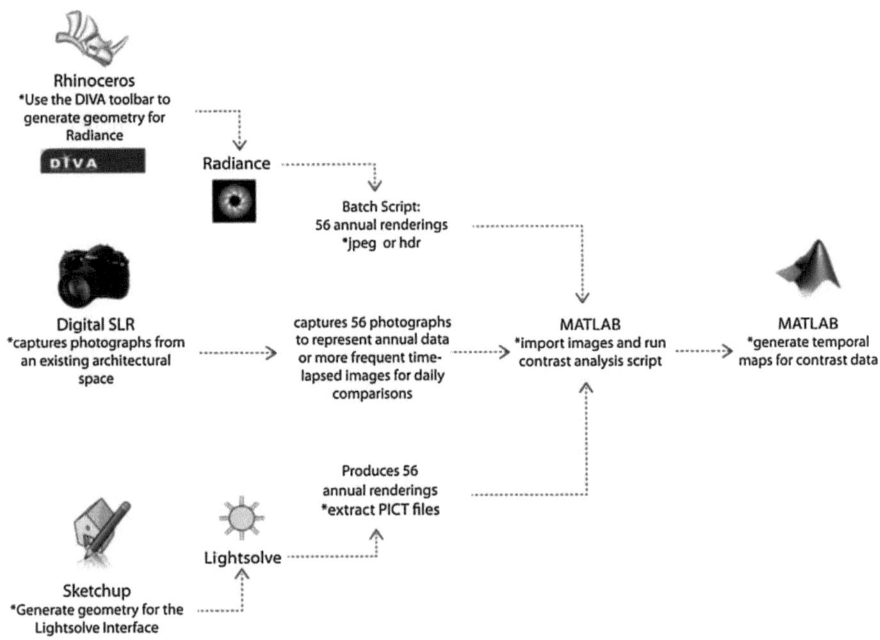

Fig. 5.2 Workflow diagram showing the potential of various modeling software packages

The second method relies on digital photographs, which can be generated from the documentation of a scaled physical model or an existing architectural space. While it can be difficult to capture 56 time-segmented photographs within a scaled model due to physical constraints (i.e., rotating the model to approximate daily and hourly sun positions), it can be even more challenging to capture time-segmented photographs within an existing space. For this, the designer must position a stationary camera, minimize sources of error such as people and moving furniture, and capture photographs using a timer. This method can be used for measuring the spatial contrast or luminance variability in an existing architectural space for a single instance or series of instances, but it is not efficient for the application of annual metrics.

To generate the renderings required for this study, we relied on Radiance, an industry standard program that runs backward ray-tracing to produce visually accurate climate-based renderings (Ward 1994). With the recent development in

DIVA, a daylight analysis toolbar developed at the Harvard Graduate School of Design (http://www.diva-for-rhino.com, 2009), it is now possible to export geometry from Rhinoceros 4.0 (http://www.rhino3d.com, 2007) directly to Radiance for analysis. This method was applied to the time segmentation logic established by Lightsolve to generate the 56 annual renderings required for the analysis of annual spatial contrast and annual luminance variability.

5.2 Modeling Assumptions

In order to produce a set of annual renderings that could be reasonably compared across all ten case studies, we modeled each space in Rhinoceros 4.0 with consistent parameters for the floor area, ceiling height, and camera location (Fig. 5.3). The camera was positioned to face south and was centered in the east–west direction, offset ten feet from the rear wall (Fig. 5.4), to ensure an even distribution of wall, floor, and ceiling surfaces within each view. The DIVA for Rhinoceros toolbar was then used to export the camera view to Radiance with a vertical and horizontal viewport ratio set to −vv 40 and −vh 60. The specified materials for each surface were set to default reflectance values for floor, wall, and ceiling surfaces (0.3, 0.7, 0.9 respectively). The resolution of each image was rendered at 'high quality' to accommodate adequate detail with a 640 × 480 pixel aspect density. Boston, Massachusetts was the selected location for all case study renderings (latitude 42 N, longitude 72 W). The exact date and time for each rendering was established by subdividing the year into 8 symmetrical dates and each date into seven symmetrical times from sunrise to sunset (Fig. 5.5), a method validated by Lightsolve (Kleindienst et al. 2008).

Although these metrics could eventually account for dominant sky conditions and evaluate the effects of climate on annual contrast, we determined that the

Fig. 5.3 Dimensions for model

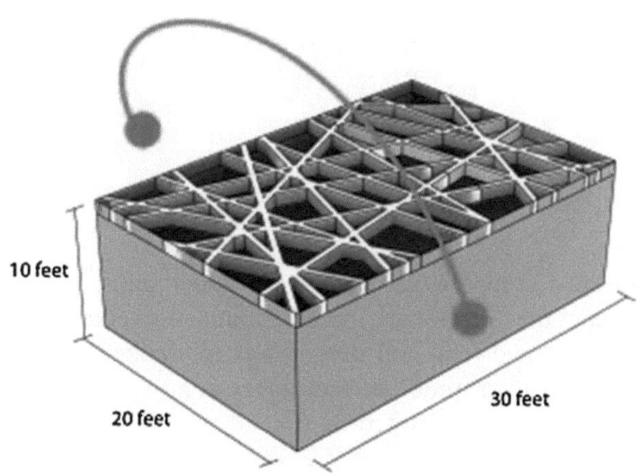

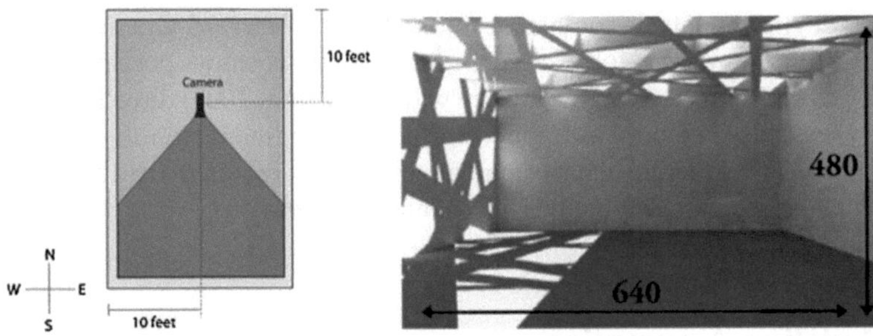

Fig. 5.4 Camera location in plan and image pixel aspect ratio

(a)

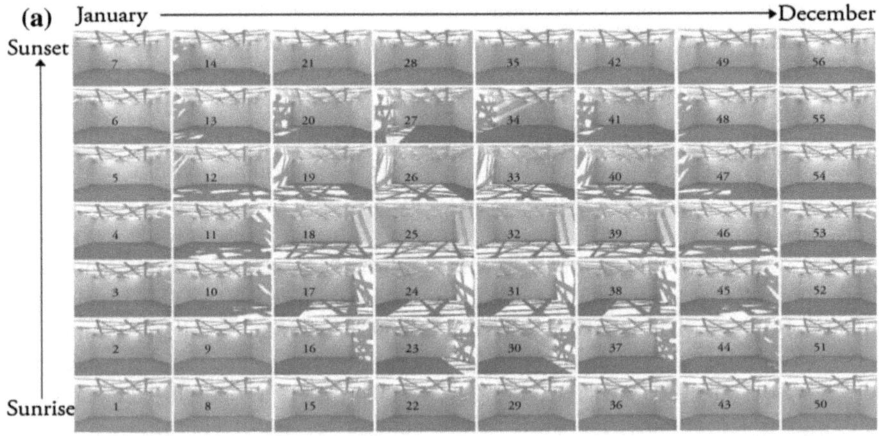

(b)

	Date	Time
1	1/13	7:52
2	1/13	9:12
3	1/13	10:33
4	1/13	11:53
5	1/13	13:13
6	1/13	14:34
7	1/13	15:54
8	2/28	7:09
9	2/28	8:45
10	2/28	10:21
11	2/28	11:57
12	2/28	13:33
13	2/28	15:09
14	2/28	16:45

	Date	Time
15	4/14	7:03
16	4/14	8:57
17	4/14	10:51
18	4/14	12:45
19	4/14	14:39
20	4/14	16:33
21	4/14	18:27
22	5/30	6:15
23	5/30	8:24
24	5/30	10:33
25	5/30	12:42
26	5/30	14:50
27	5/30	16:59
28	5/30	19:08

	Date	Time
29	7/15	6:25
30	7/15	8:33
31	7/15	10:41
32	7/15	12:50
33	7/15	14:58
34	7/15	17:06
35	7/15	19:14
36	8/29	7:03
37	8/29	8:57
38	8/29	10:51
39	8/29	12:45
40	8/29	14:39
41	8/29	16:33
42	8/29	18:27

	Date	Time
43	10/14	7:43
44	10/14	9:18
45	10/14	10:54
46	10/14	12:29
47	10/14	14:04
48	10/14	15:40
49	10/14	17:15
50	11/28	7:30
51	11/28	8:50
52	11/28	10:11
53	11/28	11:32
54	11/28	12:52
55	11/28	14:23
56	11/28	15:33

Fig. 5.5 **a** Temporal map of 56 annual instances and **b** 56 annual dates/times for Boston, MA

clearest comparisons were made under sunny skies with direct light penetration. When the spaces are rendered under overcast sky conditions, the amount of contrast and temporal diversity is minimized. In order to analyze the impacts of contrast over time, it was necessary to use a sky condition that allowed for maximized visual effects.

Images produced by Radiance are reduced from high dynamic range (HDR) to jpeg format for speed and ease of processing. Although jpeg images have a narrower range of pixel information (0–255) than that of HDR images, the compression is linear and represents a proof-of-concept within the limits of what we can observe on screen. A method that utilizes tone-mapped HDR images is necessary to measure the space with dynamics more closely related to the human eye. This will be explored in depth as the research progresses. The date and time for each rendering can be adjusted for various latitudes to accurately describe even daily subdivisions. Once we have produced these annual sets of renderings, we can generate data for annual spatial contrast and annual luminance variability and map those effects over the year to see how they are affected by dynamic sun conditions.

5.3 Case Study Results

To calculate annual spatial contrast and annual luminance variability, each set of radiance renderings is imported into MATLAB so that individual images may be processed and data may be overlaid between images. The results of these metrics can be seen in their application to each of the following four typological models: category one (*Direct and Exaggerated*), category four (*Partially Direct and Screened*), category nine (*Indirect and Dispersed*), and category ten (*Indirect and Diffuse*).

Category one, *Direct and Exaggerated*, represents a top-lit space with thickened asymmetrical mullions creating a dramatic penetration of sunlight across the walls and floor (Fig. 5.8). Category four, *Partially Direct and Screened,* displays a louvered, side-lit daylight strategy that produces high spatial contrast and luminance variability in the winter, with less dramatic effects occurring during the summer when the sun is high in the sky (Fig. 5.11). Category nine, *Indirect and Dispersed*, has a north-facing sawtooth roof that minimizes contrast and variability with a daylight strategy that allows for minimal sunlight penetration (Fig. 5.14). Category ten, *Indirect and Diffuse*, represents the low-contrast, low-variability end of the spectrum with a translucent glass roof that distributes an even and stable luminosity across the visual field (Fig. 5.17).

The numerical scale for each metric, spatial contrast and luminance variability, has been determined by the results from all ten case studies. Based on the distribution of resulting values, two thresholds divide each set of data into three parts, each representing a third of the population. As a result of this statistical subdivision, spatial contrast values between 0 and 0.5 are considered low, values between 0.5 and 0.8 are considered medium, and values exceeding 0.8 are considered high (Fig. 5.6). Luminance variability values between 0 and 2×10^6 are considered low, values between 2×10^6 and 3×10^6 are considered medium, and values exceeding 3×10^6 are considered high (Fig. 5.7). We will use these relative thresholds to discuss the results in terms of relative high, medium, and low, although future research is needed to develop a more statistically accurate range for each metric.

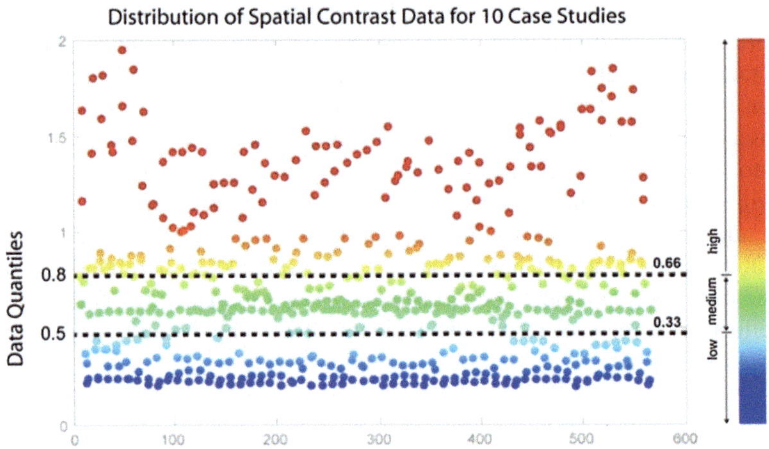

Fig. 5.6 Plot of spatial contrast values for 10 case studies. Based on the distribution of data, we have established two thresholds that evenly divide the data into three ranges: high (>0.8), medium (0.5–0.8), and low (<0.5)

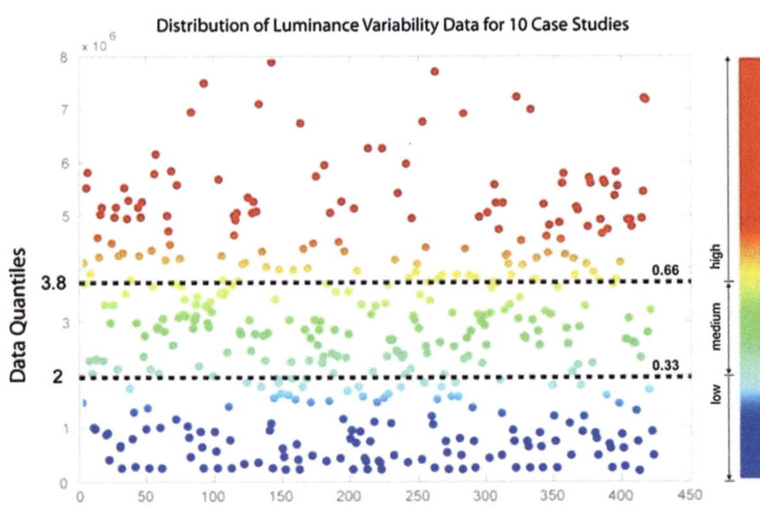

Fig. 5.7 Plot of luminance variability values for 10 case studies. Based on the distribution of data, we have established two thresholds that evenly divide the data into three ranges: high (>3.8×10^8), medium (2×10^8–3.8×10^8), and low (<2×10^8)

5.3.1 Category One, Direct and Exaggerated

Category one, *Direct and Exaggerated*, modeled to represent a highly contrasted and variable space, demonstrates a consistently high degree of spatial contrast throughout the year (Fig. 5.8). The temporal map in Fig. 5.9 shows a peak in spatial

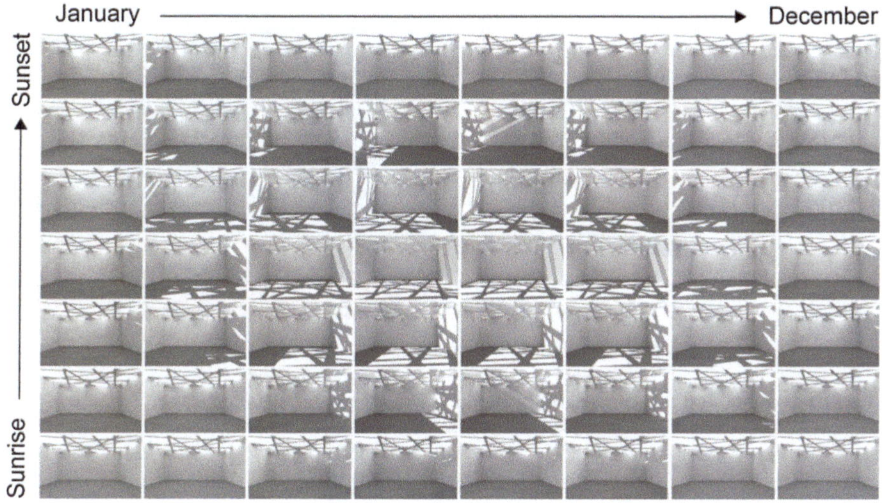

Fig. 5.8 Annual renderings for category one, *Direct and Exaggerated*

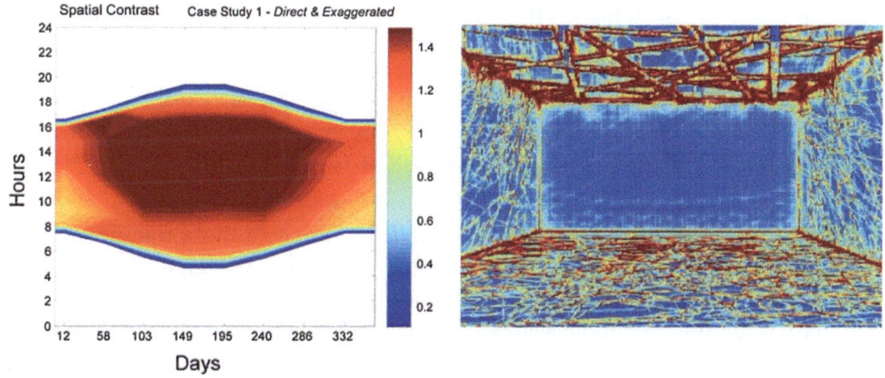

Fig. 5.9 Annual spatial contrast (temporal map and cumulative image)

contrast between 10 a.m. and 2 p.m. in the summer months when the sun is directly overhead, while the temporal map of luminance variability in Fig. 5.10 shows maximum variations occurring throughout the spring and fall. These variations in luminance are due to the changing altitude of the sun, which causes fluctuations in brightness throughout the space. In the image to the right of Fig. 5.9, thick red lines signify where spatial contrast is most consistent, highlighting the roof structure as the most redundant source with secondary accumulations on the floor and walls. The image to the right of Fig. 5.10 depicts a cumulative view of annual luminance variation. Here, the floor is the area that experiences the most dramatic change throughout the year. The cumulative effects shown in these two false-color images present an important distinction between metrics. Annual spatial contrast shows

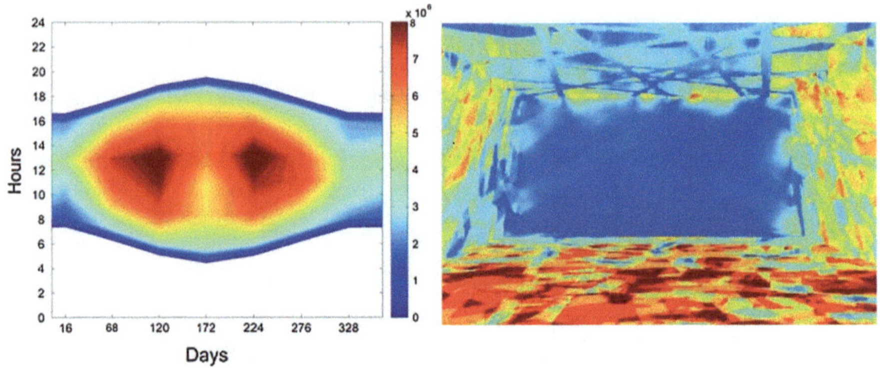

Fig. 5.10 Annual luminance variability (temporal map and cumulative image)

areas within the view where contrast is accumulated, highlighting redundant textures and forms, whereas annual luminance variability shows areas within the image that experience the most variation, emphasizing areas of instability. When compared side by side, these metrics allow us to discuss spatial contrast and temporal variability as related, but distinct visual characteristics present within the architectural space.

5.3.2 Category Four, Partially Direct and Screened

Category four, *Partially Direct and Screened*, represents a more traditional side-lit daylight strategy with a clerestory window above and louvered screen below, creating varied effects across the year depending on solar altitude (Fig. 5.11). This case study is reminiscent of the Magney House, designed by Glen Murcutt. Here, the results for annual spatial contrast and luminance variability depict more temporal variation, with a dramatic shift in contrast between the winter and summer months.

The temporal map in Fig. 5.12 shows high spatial contrast between October and February, with medium contrast throughout the rest of the year. The location of these effects can be seen in the false-color image to the right of Fig. 5.12, which shows the accumulation of contrast on the walls and floor closest to the wall of louvers.

Annual luminance variability, as seen in the temporal map in Fig. 5.13, shows similar changes in strength across the winter and summer months, with a concentration of these changes at sunrise and sunset, when the angle of sunlight allows for deep penetration within the space. These variations range from low to high variability and occur most frequently on the walls and floor adjacent to the louvers.

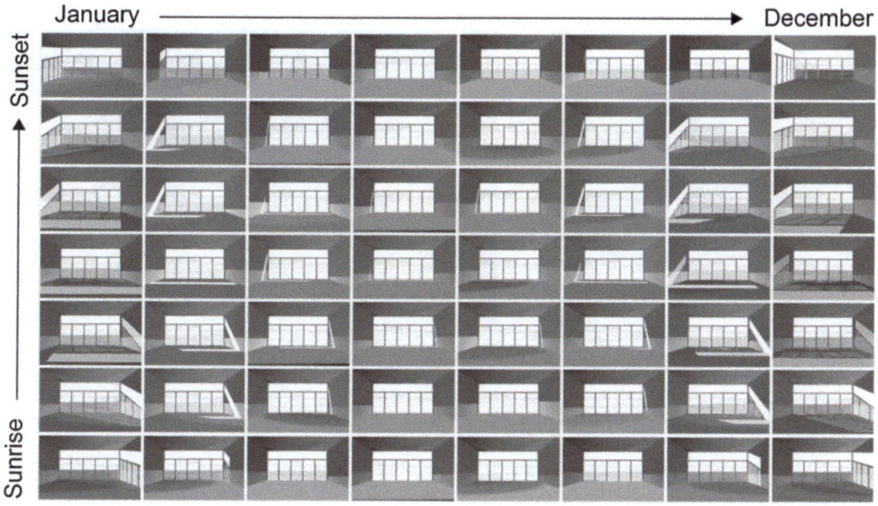

Fig. 5.11 Annual renderings for category four, *Partially Direct and Screened*

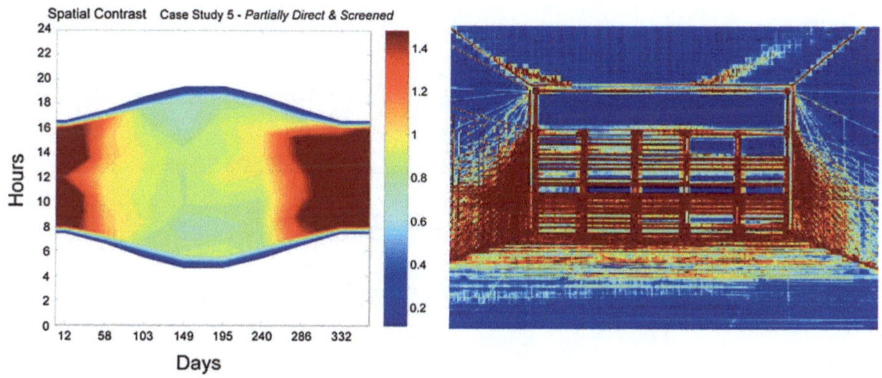

Fig. 5.12 Annual spatial contrast for category four (temporal map and cumulative image)

5.3.3 Case Study Space Nine, Indirect and Dispersed

Category nine, *Indirect and Dispersed*, contains a series of north-facing roof monitors that emit diffuse daylight down into the space. This case study was inspired by the Dia Beacon Museum in upstate New York, designed by Robert Irwin and Open Office. Across most of the day and year, case study nine achieves uniform lighting levels; however, there are moments of sharp variability that occur as sunlight penetrates the roof monitors in the early morning and late afternoon, as seen through the renderings in Fig. 5.14.

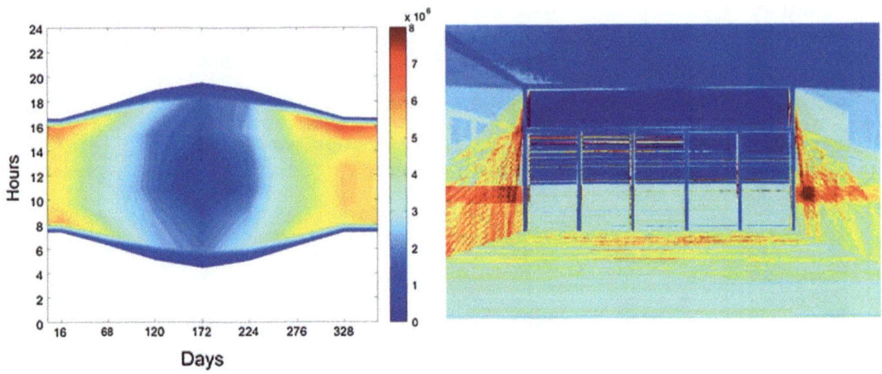

Fig. 5.13 Annual luminance variability for category four (temporal map and cumulative image)

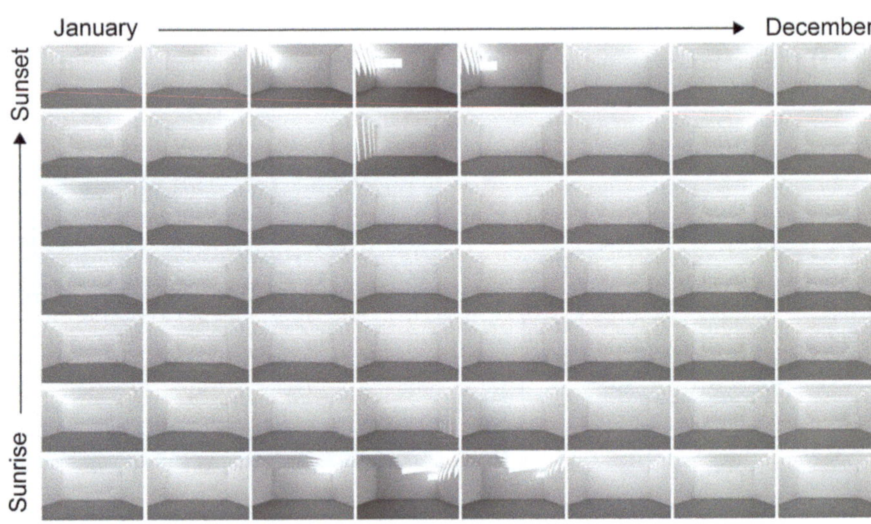

Fig. 5.14 Annual renderings for category nine, *Indirect and Dispersed*

The temporal map in Fig. 5.15 shows low spatial contrast throughout the year with slight variations at sunrise and sunset during the summer months. Luminance variability, however, is much more dynamic. It ranges from high in the early mornings to low at noon and then back to high just before sunset (Fig. 5.16). This shift is due to low solar altitudes in the morning and late afternoon, which allows direct sunlight to penetrate the skylights and cast shadows across the walls and floor. While this causes minimal spatial contrast throughout the year, it does show a dramatic impact on luminance variability at sunrise and sunset throughout the spring and summer months. The image to the right in Fig. 5.15 shows the location of spatial contrast along the roof monitors, while the image to the right in Fig. 5.16

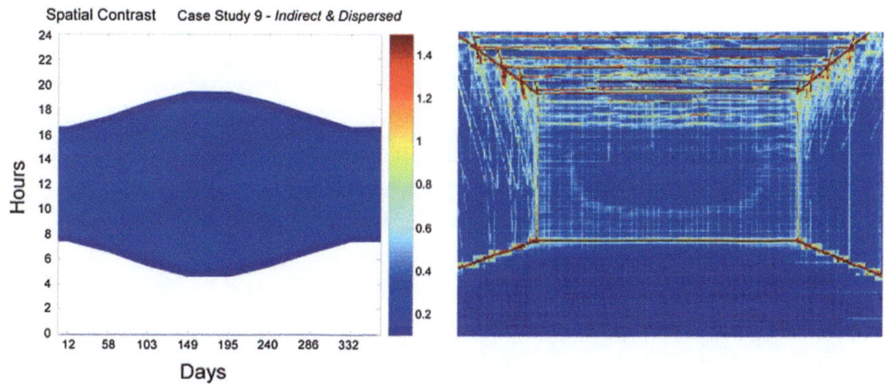

Fig. 5.15 Annual spatial contrast (temporal map and cumulative image)

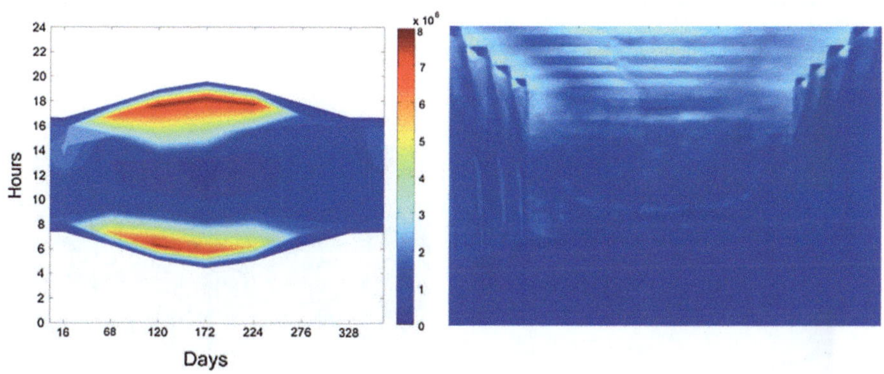

Fig. 5.16 Annual luminance variability (temporal map and cumulative image)

shows minimal luminance variability on the floor and walls, with a moderate degree occurring across the ceiling.

5.3.4 Category Ten, Indirect and Diffuse

Category ten, *Indirect and Diffuse,* shows a space with very little luminance variability and minimal spatial contrast. This space was modeled after the Modern Wing at the Chicago Art Institute, designed by Renzo Piano. The translucent glazed roof diffuses incoming sunlight, creating a uniform distribution of daylight that can be seen throughout the annual renderings (Fig. 5.17).

The dark shadows produced by the back-lit mullions generate some spatial contrast across the ceiling, evident in the image to the right of Fig. 5.18. The temporal

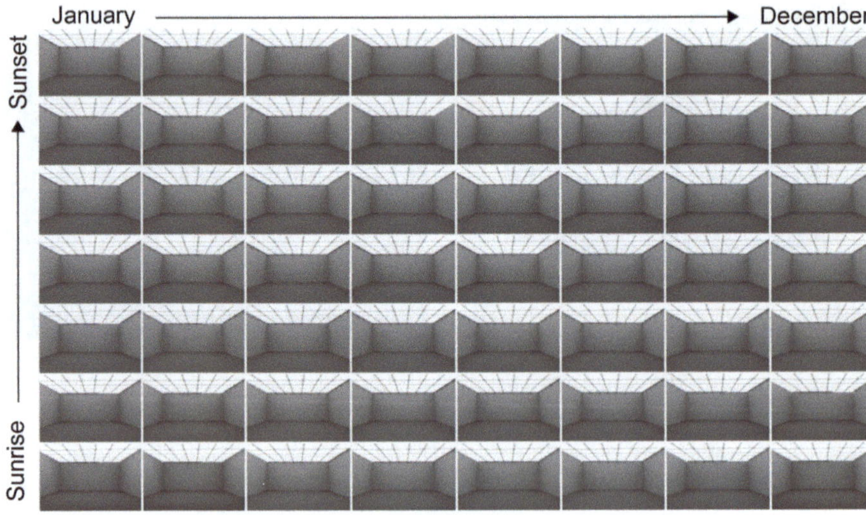

Fig. 5.17 Annual renderings for category ten, *Indirect and Diffuse*

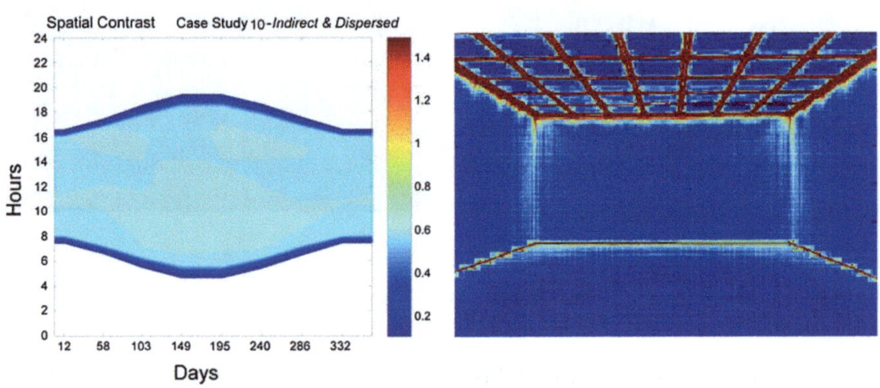

Fig. 5.18 Annual spatial contrast (temporal map and cumulative image)

map to its left, however, displays this spatial contrast as consistently low to medium throughout the year, with minimal daily or seasonal variations. The temporal map in Fig. 5.19 shows minimal luminance variability throughout the year.

It is important to mention that this particular space was simulated in two separate attempts with the first set of images representing sharp contrast between the mullions and glass, with peaks of high variability as the sun altered those shadows across the translucent glazing. This raises an important potential error within the production of rendered images, which must represent an accurate view of the interior space. Perceptual field-of-view metrics cannot calculate the degree of perceived spatial contrast or luminance variability without an accurate set of images.

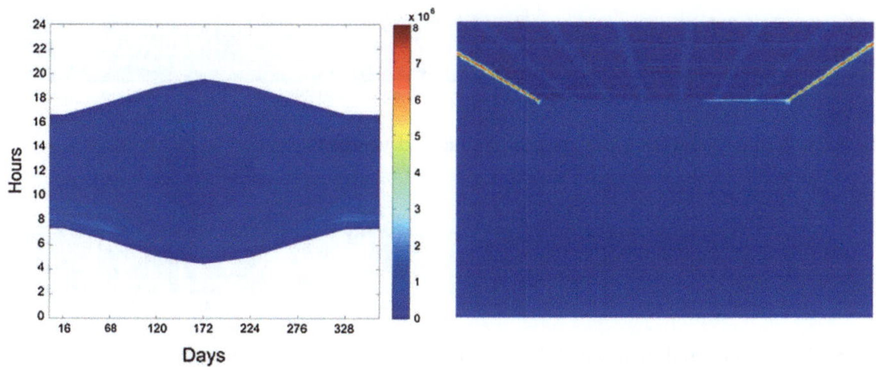

Fig. 5.19 Annual luminance variability (temporal map and cumulative image)

5.4 Assessing Results for the Case Study Spaces

From this set of results, a clear gradient can be identified between the high and low ends of the contrast and variability spectrum. The four examples presented in Sect. 5.2 represent various combinations of annual spatial contrast and luminance variability, strengthening the need for both metrics to depict the complexity of temporal conditions within each space. The results presented in this chapter are conclusive in their quantitative methods for describing each effect; however, they raise the question of how these numbers can or should be combined to represent an overall ranking. Of the original fifteen categories described in the typological matrix, only ten were modeled and analyzed. Figure 5.20 shows these ten case studies with high contrast and variability on the left to low contrast and variability on the right, which was intuitively established before any quantitative analyses. In order to compare the annual results of each metric, dynamic data for each of the 56 instances were compressed into a single number that represents a cumulative sum of results. For annual spatial contrast, this number represents the sum of instantaneous spatial contrast across all 56 images. For annual luminance variability, this number represents a cumulative sum of variation in brightness across the same set of images.

Figure 5.21 shows the ten case studies reconfigured to represent a new gradient specific to annual spatial contrast, while Fig. 5.22 shows the case studies reconfigured for annual luminance variability. These linear gradients organize each case

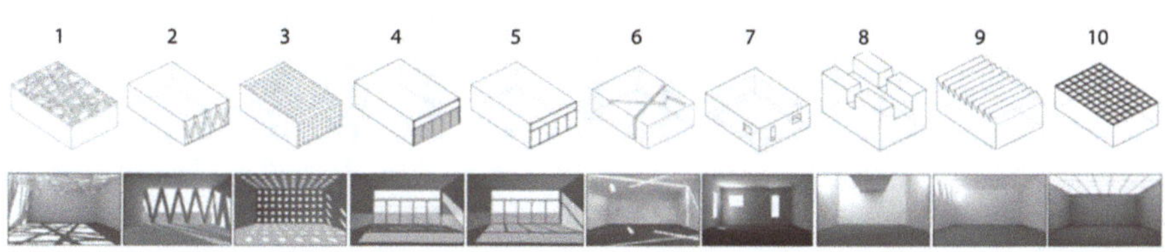

Fig. 5.20 Original gradient of hypothesized effects

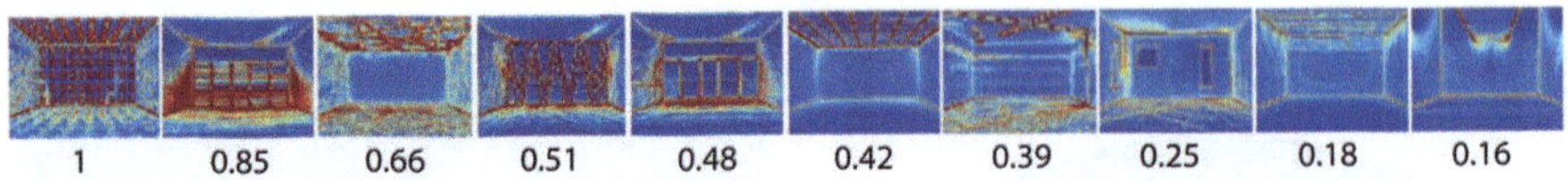

Fig. 5.21 Reconfigured matrix based on normalized values for annual spatial contrast

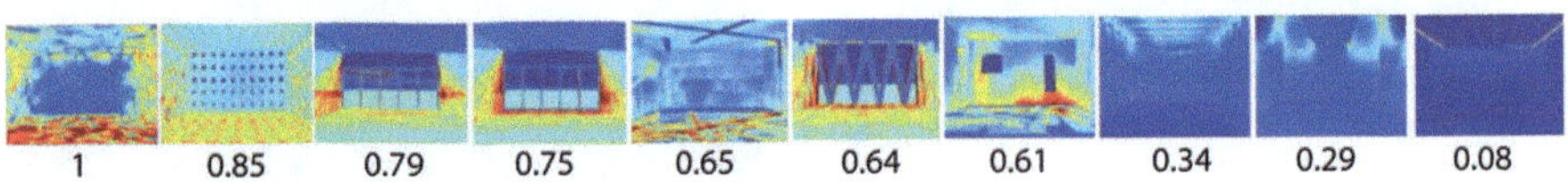

Fig. 5.22 Reconfigured matrix based on normalized valued for annual luminance variability

study into a line from high annual spatial contrast and luminance variability on the left to low annual spatial contrast and luminance variability on the right. A temporal plot of these results can be seen in Fig. 5.24 (annual spatial contrast) and Fig. 5.25 (annual luminance variability). To compare the new metrics side by side, the resulting values for annual spatial contrast and annual luminance variability have been normalized. Figure 5.23 shows a mostly linear trend in results for each annual metric. While case studies 1 and 10 break the linear trend for annual spatial contrast, they fit the trend for annual luminance variability. Case studies 2 and 9, on the other hand, appear as outliers for both annual metrics. Where case study 2 appears to be located in the wrong position within the intuitive gradient (we feel that it should have been located closer to the middle after reviewing the annual renderings), the metrics revealed surprising changes in luminance and contrast in case study 9. These changes were difficult to anticipate due to the geometry of the roof and incident solar angles, and we believe that the discrepancy between our

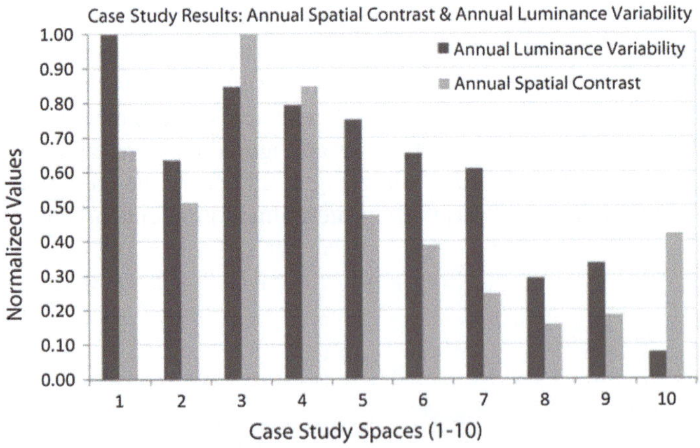

Fig. 5.23 Normalized values for annual spatial contrast and annual luminance variability for each of the ten case study spaces

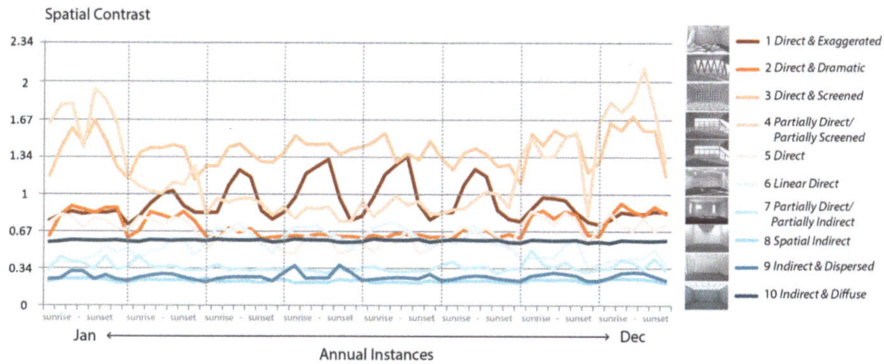

Fig. 5.24 Temporal plot of spatial contrast values for all 10 case studies

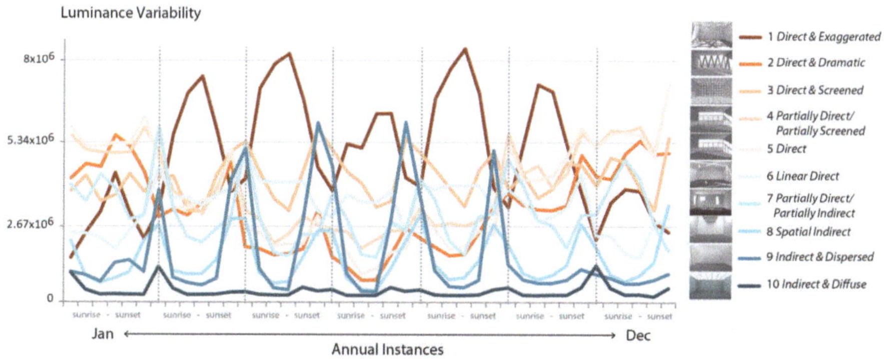

Fig. 5.25 Temporal plot of luminance variability values for all 10 case studies

intuition and the proposed metrics further exemplifies the importance of dynamic visual analysis methods. In this case, the metrics helped to reveal perceptual changes within the visual field through a temporal analysis.

In order to compare or combine these effects to recreate the original hypothesized matrix (which represented a combination of the two annual metrics), one would need to validate the study across a wider range of typological examples to establish a more statistically accurate scale for each metric, allowing for a weighted adjustment to the cumulative annual values. This would allow the two annual metrics to be combined and compared on a relative scale from high to low.

This pre-validation study does, however, allow us to make comparisons between proposed metrics for each of the ten case study spaces. For example, we can say that the louvered space in case study 6 represents a relatively high degree of both annual spatial contrast and annual luminance variability, while the screened space in case study 4 shows a high degree of annual spatial contrast with a low degree of annual luminance variability. These numbers also allow us to confirm that case study 1 experiences more annual luminance variability than any other space within the study. Although the numerical representation for each metric requires further development, these pre-validation results demonstrate the capabilities of annual spatial contrast and annual luminance variability in quantifying a set of temporal characteristics within daylit architecture.

References

http://www.diva-for-rhino.com. (2009). Retrieved from DIVA-for-Rhino.
http://www.rhino3d.com. (2007). Retrieved 2010, from Rhinoceros.
Kleindienst, S., Bodart, M., & Andersen, M. (2008). Graphical representation of climate based daylight performance to support architectural design. *LEUKOS, 5*(1), 39–61.
Ward, G. (1994). The RADIANCE lighting simulation and rendering system. In *Proceedings of '94 SIGGRAPH Conference*, (pp. 459–472).

Chapter 6
Application of New Metrics to Detailed Case Studies

Keywords Spatial contrast • Annual spatial contrast • Annual luminance variability • Architectural lighting design • Daylight analysis • Daylight simulation

In the previous chapter, we applied spatial contrast, annual spatial contrast, and annual luminance variability metrics to a series of rendered case study spaces. The results for each metric were compared to show how they differentiate between dynamic qualities of contrast and luminous diversity within each case study space. In this chapter, we will now apply annual spatial contrast and luminance variability metrics to rendered models of two existing architectural spaces: Toyo Ito's 2002 Serpentine Pavilion and Louis Kahn's First Unitarian Church. The results will be discussed alongside existing metrics such as Daylight Factor, Daylight Autonomy, and Daylight Glare Probability to provide a more holistic assessment of daylight performance within each space.

6.1 Modeling Assumptions

Each architectural space was modeled in Rhinoceros (http://www.rhino3d.com) and assigned default radiance materials for floor, wall, and ceiling surfaces (0.3, 0.7, and 0.9 respectively). Although the geometry for each space was modeled accurately from existing documentation, default reflectance values were used because detailed material properties were unknown for a majority of surfaces. The camera view was selected to mimic an existing photograph of each space so that interior lighting conditions would be adequately represented from a human perspective. The location of each model was adjusted in Radiance, with Kahn's Church located in Rochester, NY (43 N, 77 W) and Ito's Pavilion in London, UK (51 N, 10 W). The rendering quality, view aspect ratio, and pixel resolution were set to high-quality ray-tracing, 40 × 60, and 480 × 640, respectively.

S. Rockcastle and M. Andersen, *Annual Dynamics of Daylight Variability and Contrast*, SpringerBriefs in Computer Science,
DOI: 10.1007/978-1-4471-5233-0_6, © The Author(s) 2013

6.2 2002 Serpentine Pavilion

Toyo Ito's Serpentine Pavilion, constructed in Hide Park, London in 2002, was part of an ongoing commission instituted by the Serpentine Gallery for the exhibition of contemporary architecture. Each year, a prominent architect is selected to design and build a pavilion in which public activities such as film screenings, receptions, and lectures can take place. The design for the 2002 pavilion was a box of diagonal steel members with alternating glass and opaque white panels. The overall dimensions for the structure were 60 feet × 60 feet × 15 feet and were modeled from existing plan and elevation drawings of the building (Fig. 6.1). The primary structure for the pavilion relied on a web of intersecting steel members (Fig. 6.1b), with secondary panels providing shear support (Fig. 6.2b). The remaining openings were covered in glass to provide protection from the elements, while maintaining transparency to the exterior (Fig. 6.2a). According to our contrast and variability matrix (presented in Chap. 3), this space would fall into category one and represent a *Direct and Exaggerated* daylight strategy. Due to the temporary public program of this pavilion and its use as a semi-outdoor venue, there was little need to minimize direct sunlight within the space which became exaggerated through the asymmetry and transparency roof and wall elements.

The photograph in Fig. 6.3a shows the southeast corner of the pavilion while the rendering in Fig. 6.3b shows the selected camera angle, at approximately the same location where the photograph was taken. The annual set of renderings in Fig. 6.4 shows a variable space with large patches of direct sunlight casting dynamic shadows across the floor and walls. The temporal maps and cumulative false-color images on the opposite page show the magnitude of this contrast and variation in luminance across the year. The temporal map in Fig. 6.5 shows high spatial contrast throughout the year, with a concentration from 10 a.m. to 4 p.m. during the summer months. The cumulative image to the right shows where this contrast occurs most frequently, highlighting lines of structure in the roof and resulting patterns across the floor. Figure 6.6 shows a dynamic temporal map with

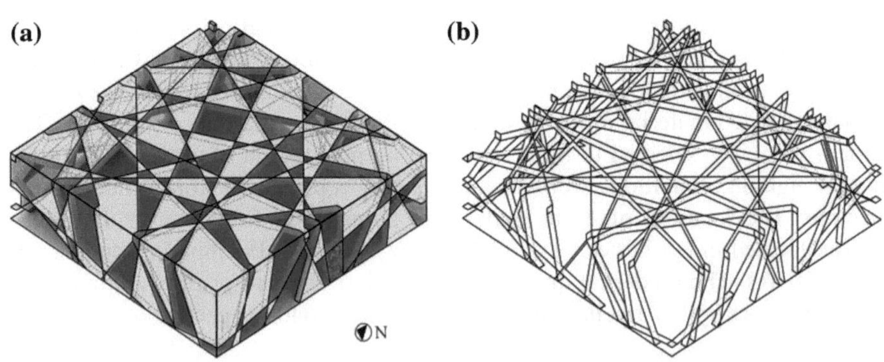

Fig. 6.1 **a** Axon of Serpentine Pavilion facing northeast, **b** structural steel members

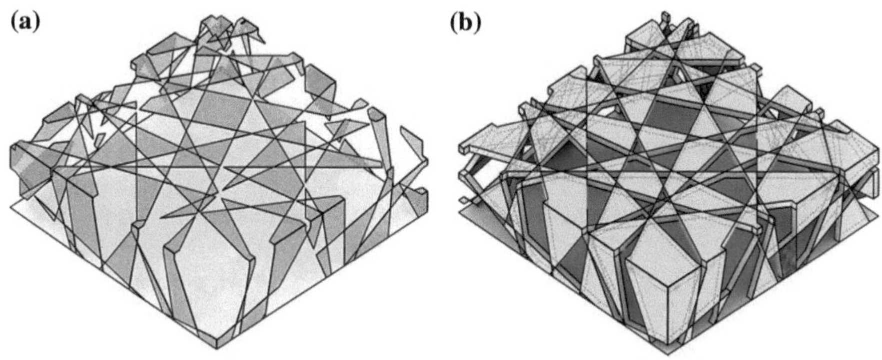

Fig. 6.2 Axon of Serpentine Pavilion showing glazing (**a**), panels, and structure (**b**)

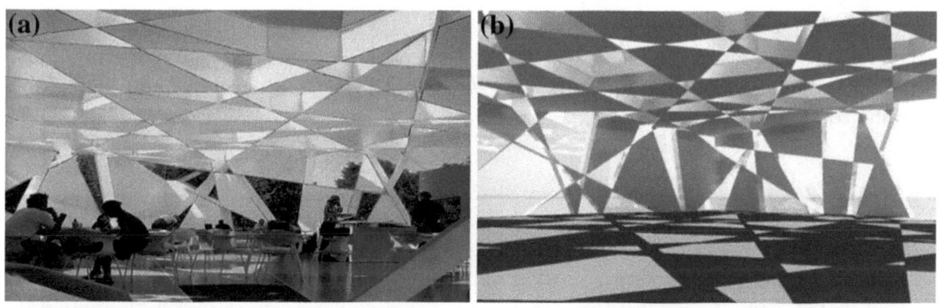

Fig. 6.3 Serpentine Pavilion: **a** Photograph, nclave, May 26, 2007 via flickr, Creative Commons License, and **b** Rendering of interior space, DIVA for Rhinoceros, http://www.diva-for-rhino.com/

peaks in luminance variability during the summer months and various degrees of change occurring throughout the rest of the year. The accumulative image to the right shows these variations occurring most frequently across the floor.

Figure 6.7 shows a base-line analysis across all 56 images to differentiate between accumulative brightness and the variation that is revealed by each of the annual metrics presented by this book. 'Annual Accumulative Brightness,' as we will call it, takes the sum of all pixel values within an image and plots them across the temporal map to the left of Fig. 6.7. The image to the right shows a simple accumulation of all 56 renderings, highlighting areas that are consistently bright. The temporal map on the left and the image on the right show how brightness accumulates across the year, while the metrics represented by Figs. 6.5 and 6.6 show how it *varies*, exposing the dynamic nature of daylight. Annual spatial contrast and luminance variability add more depth to our understanding of architecture over time.

The need for these visually dynamic annual metrics emerged out of a critical analysis of existing daylight metrics such as Daylight Factor (DF), Daylight

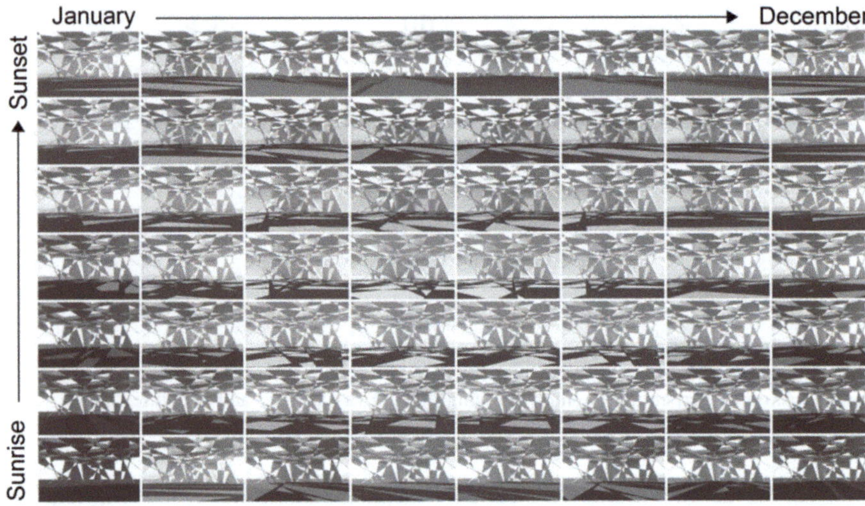

Fig. 6.4 Annual renderings of the Serpentine Pavilion, DIVA for Rhinoceros, http://www.diva-for-rhino.com/

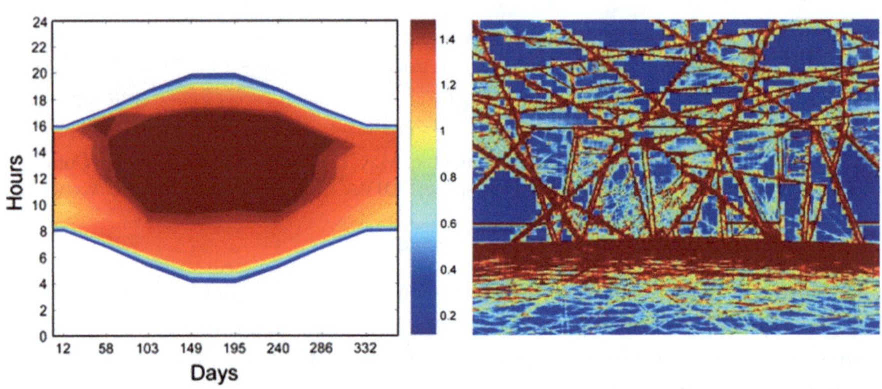

Fig. 6.5 Annual spatial contrast for the Serpentine Pavilion (temporal map and cumulative image)

Autonomy (DA), and Daylight Glare Probability (DGP) and their inability to capture the spatial and temporal diversity of daylight within our field of view. In order to differentiate the contrast-based metrics proposed by this research, DF, DA, and DGP analyses were run on both the Serpentine Pavilion and the First Unitarian Church to expose their limitations in describing temporal visual qualities of architecture and show how new metrics can be used as a compliment.

A daylight factor analysis conducted in DIVA (Fig. 6.8a), with a sensor grid 2.5 feet from the floor of the pavilion (standard table height), shows that there is more than sufficient illumination for occupants to perform basic tasks such as

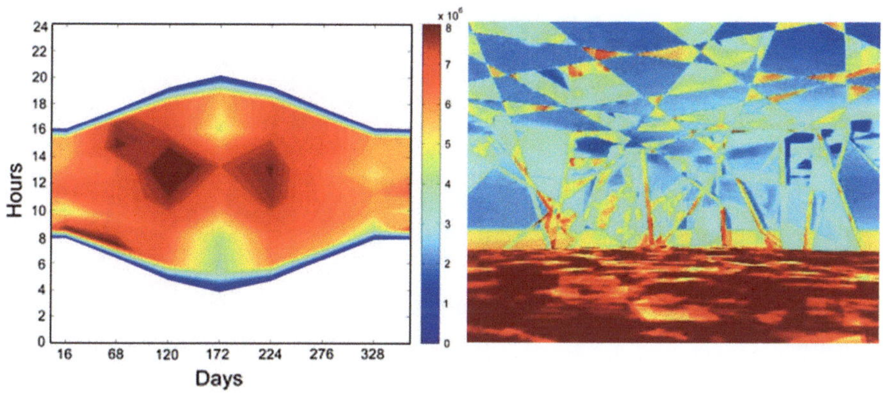

Fig. 6.6 Annual luminance variability for the Serpentine Pavilion (temporal map and cumulative image)

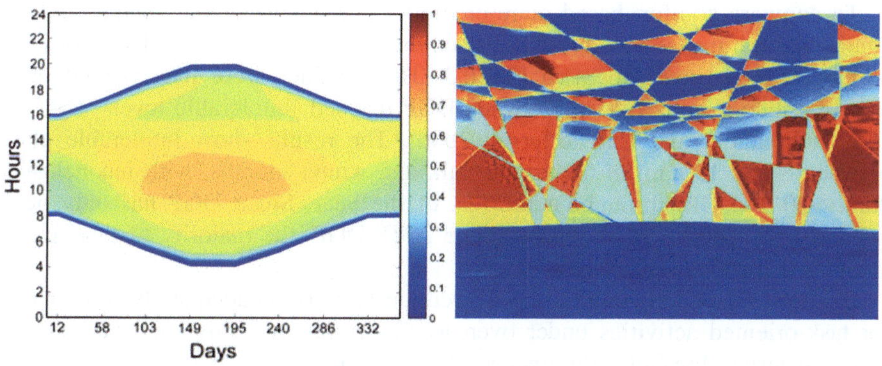

Fig. 6.7 Annual luminance accumulation for the Serpentine Pavilion (temporal map and cumulative image)

reading and writing (2–5 %) under overcast sky conditions (BS8206-2 2008). With a mean DF of 13.12 %, we can infer that there will be problems with heat gain due to excess illumination, although we cannot infer the added impacts of direct sunlight and dynamic sky conditions which should further amplify the problems. Daylight autonomy (Fig. 6.8b) shows a more comprehensive analysis of illumination for the task plane (Reinhart and Walkenhorst 2001). With a minimum threshold of 300 lux and occupancy hours from 8 a.m. to 6 p.m. (365 days a year), the mean DA is 89 % with 97 % of the space achieving a DA of 50 % or higher. While DA shows that we should have *enough* light for task performance, a useful daylight illuminance UDI simulation, run through DAYSIM shows that only 7 % of the space achieves a $\text{UDI}_{<100\text{–}2000\ \text{lux}}$ larger than 50 % (Nabil and Mardaljevic 2006). This metric suggests that the interior of the Serpentine Pavilion receives *too much* light throughout much of the day and year.

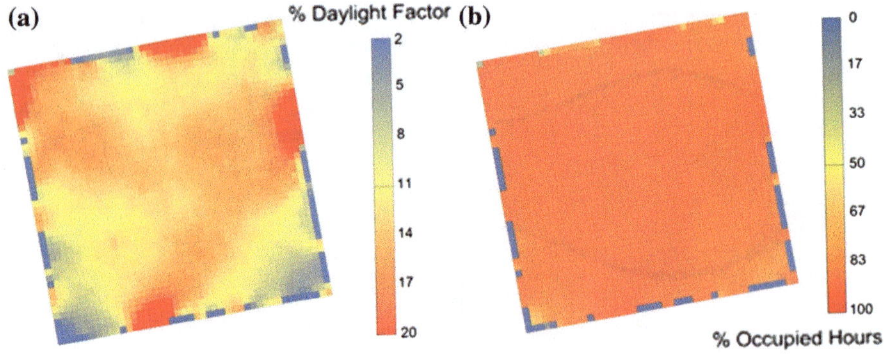

Fig. 6.8 The Serpentine Pavilion **a** Daylight Factor (2–20 %), mean DF = 13.12 %, DIVA for Rhinoceros, **b** Daylight Autonomy at 300 lux, mean DA = 86.24 % of occupied time (8 a.m.–6 p.m., 365 days a year), DIVA for Rhinoceros & DAYSIM

To compare comfort-based glare-prediction metrics to annual spatial contrast and annual luminance variability, we used the DIVA toolbar to run DGP for each of the 56 renderings shown in Fig. 6.3. These values were then plotted on a temporal map (Fig. 6.9) to show when DGP reached 'intolerable levels' (>45 % DGP) (Wienhold and Christofferson 2006). The results show intolerable glare between 8 and 10 a.m. in the spring, fall, and winter months, with inconclusive data (>20 % DGP) throughout the rest of the year. Since DGP has only been validated for side-lit office spaces above 20 % DGP, the majority of this annual analysis is inconclusive at best.

Daylight factor tells us that the Serpentine Pavilion is adequately illuminated for task-oriented activities under overcast sky conditions, DA confirms that we meet our target threshold illuminance (300 lux) throughout most of the year, and

Fig. 6.9 Serpentine Pavilion: temporal map of DGP (0–100 %), calculated for each date/time in DIVA for Rhinoceros and plotted using MATLAB

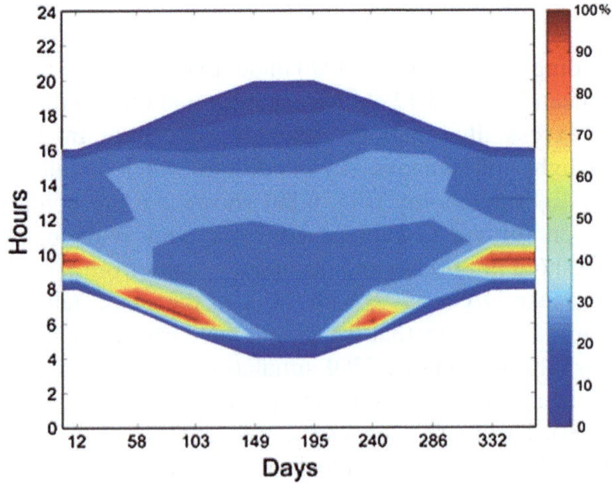

UDI tells us that we exceed the upper illuminance threshold (2,000 lux) for recommended task-based activities. The design of the Serpentine Pavilion, which is intended for temporary occupation and visually enhanced by the dramatic penetration of light and shadow cannot, however, be analyzed by task-driven illumination metrics such as DF and DA and task-driven comfort metrics like DGP. New perceptually driven metrics such as annual spatial contrast and annual luminance variability are more appropriate for visualizing and measuring the dynamic effects of sunlight in architecture. In spaces where task-activities are performed, perceptually driven metrics can be combined with task and comfort-based metrics to provide a more holistic analysis of daylight performance.

6.3 First Unitarian Church

Louis Kahn's First Unitarian Church was built in Rochester, NY in 1967. His intention was to design a space that represented the ideals of the United Universalists through essential qualities in material, structure, and light. In the brochure distributed to visitors of the church, Kahn was said to have designed the space to express 'only what matters,' with a central sanctuary surrounded by rooms devoted to education and spiritual inquiry (France 2011). The concept for the plan was based on a question mark, with the center sanctuary surrounded by layers of circulation that allow for various degrees of separation. This is achieved through a multilayered box, with internal and external concrete walls (Fig. 6.10). The inner layer, a 15 ft. concrete masonry block wall, supports four branching concrete columns, which in turn carry the structure of the roof (Fig. 6.10). The outer layer of the sanctuary is constructed from cast-in-place concrete and terminates in four 30-ft-high roof monitors with internally facing clerestory windows. These roof monitors emit mostly indirect light, which bounces off the outer concrete wall and down into the sanctuary. This creates smooth gradients of light in all four corners of the church. There is some spatial contrast present within the space, but the dominant visual effects are slowly changing luminance levels across the year.

Indirect light is emitted to the central sanctuary as a smooth gradient across the outer and inner concrete walls. The photograph in Fig. 6.11a captures the northeast corner of the inner sanctuary, while the rendering in Fig. 6.11b shows the orientation of the selected camera angle, set at approximately the same location. The annual renderings, as seen in Fig. 6.12, show relatively little direct sunlight, with bands that penetrate the northeast roof monitor in the afternoon and are most pronounced during the summer months. Despite the lack of direct sunlight, there appear to be moderate fluctuations in brightness throughout the year which should impact the annual luminance variability.

Figure 6.13 shows spatial contrast values as low to medium throughout the year, with lines of contrast accentuating the concrete roof structure and expansion joints along the inner wall. As the sun moves across the vertical monitors, large shifts in overall brightness create a dynamic temporal map for luminance variability.

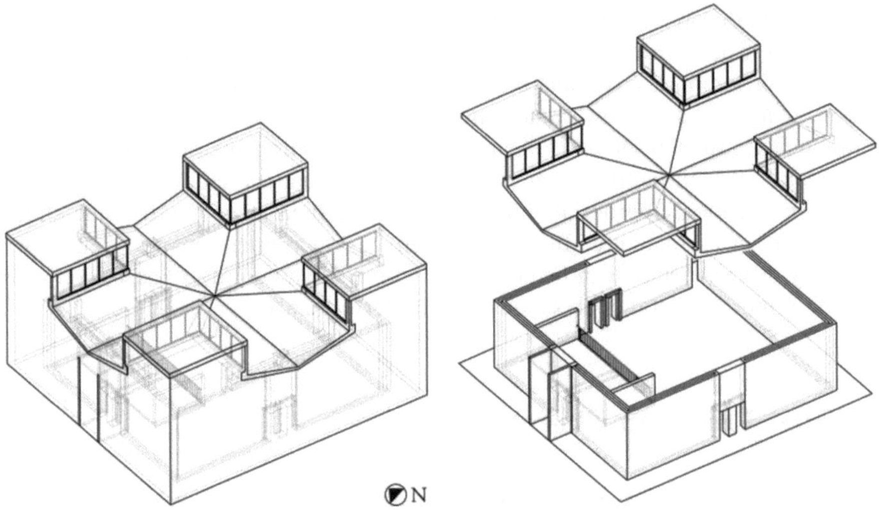

Fig. 6.10 Axon of First Unitarian Church looking north

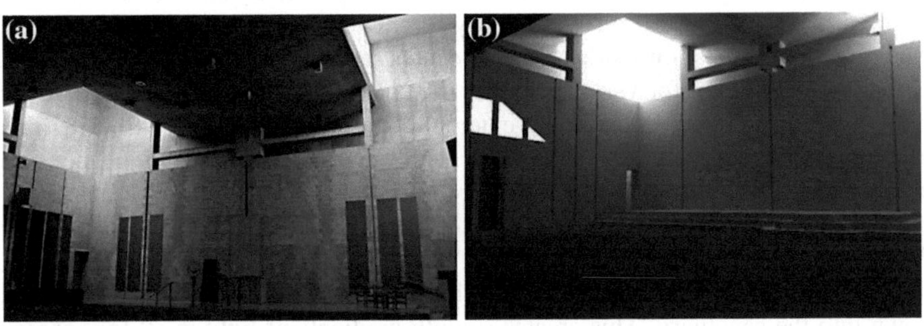

Fig. 6.11 First Unitarian Church **a** Photograph © Bryan Maddock and **b** Rendering with the same approximate camera angle, DIVA for Rhinoceros, http://www.diva-for-rhino.com/

This can be seen in Fig. 6.14, which shows high variability in the late afternoon, mid-morning, and throughout the day in the summer months. This temporal map is particularly engaging as it shows a wide range of luminous diversity within the church, while maintaining a relatively low degree of spatial contrast.

As a baseline comparison, Fig. 6.15 shows the accumulation of luminance levels across the 56 annual images, highlighting the corner roof monitor as a zone of brightness. When compared to the false-color images in Figs. 6.13 and 6.14, the cumulative luminance representation shows us nothing about the diversity of temporal conditions within the space. Accumulative luminance or brightness can only show us where the space is bright and when the overall brightness is comparatively higher or lower.

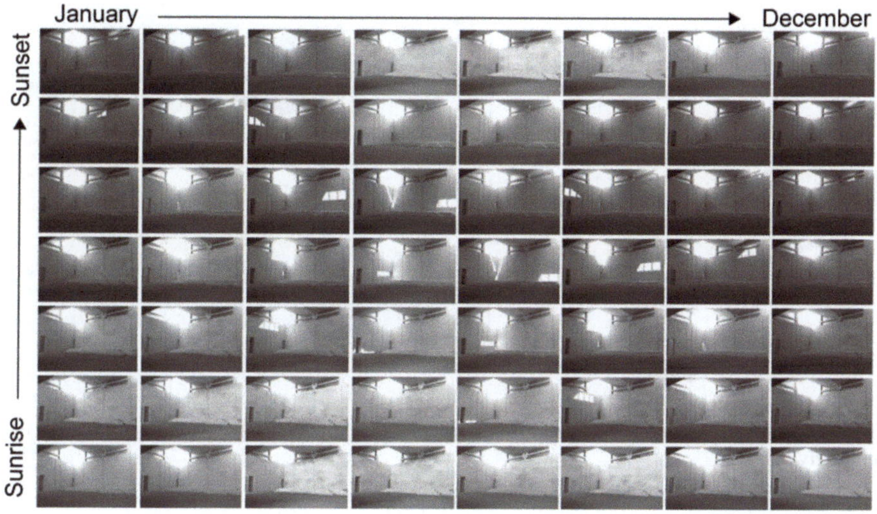

Fig. 6.12 Annual renderings for the First Unitarian Church, DIVA for Rhinoceros, http://www.diva-for-rhino.com/

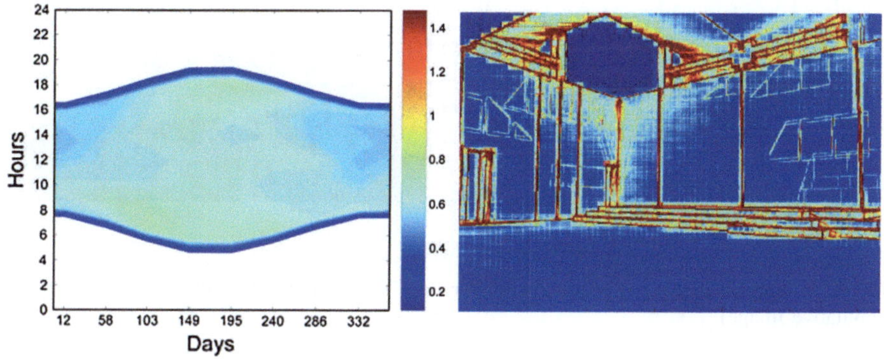

Fig. 6.13 Annual spatial contrast for the First Unitarian Church (temporal map and cumulative image)

Figure 6.16a shows a daylight factor analysis with a sensor grid 2.5 feet from the floor of the pavilion (standard table height). The mean DF across the analyzed sensor plane was 0.2 %, well below the preferred threshold for reading and writing (2–5 %) under overcast sky conditions (BS8206-2 2008). The inner sanctuary receives some task illumination, but would not meet the recommendations set by most codes and standards. Figure 6.16b shows a daylight autonomy analysis with a minimum illuminance threshold of 300 lux and occupancy hours from 8 a.m. to 6 p.m. (365 days a year) (Reinhart and Walkenhorst 2001). The mean DA, as

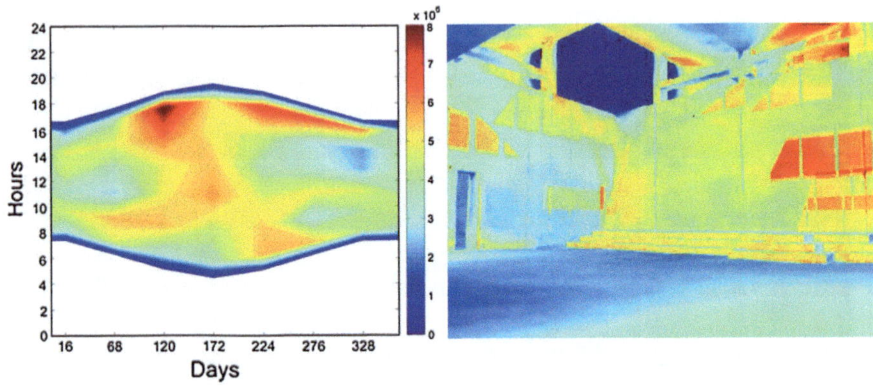

Fig. 6.14 Annual luminance variability for the First Unitarian Church (temporal map and cumulative image)

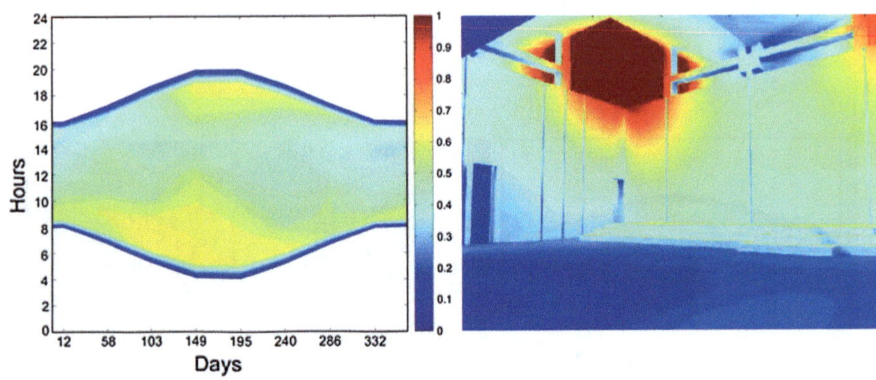

Fig. 6.15 Annual luminance accumulation for the First Unitarian Church (temporal map and cumulative image)

measured across the sensor place, is 10 % with 0 % of the space achieving a DA of 50 % or higher. The analysis map shows that only the northeast and northwest corners receive consistent illumination throughout the year, while the vast majority of the space is inadequately lit.

A DGP analysis was run for each of the 56 annual images presented in Fig. 6.12 with a temporal map of the results shown in Fig. 6.17. All the DGP results from this analysis fall between 10 and 30 % and must be interpreted carefully, as DGP values below 20 % have not been fully validated (Wienhold and Christofferson 2006). In either case, the analysis tells us that there is a low probability for task-related glare within the space.

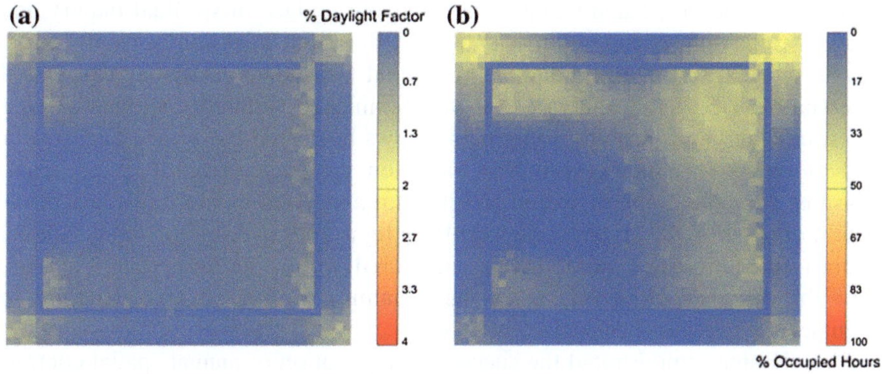

Fig. 6.16 The First Unitarian Church **a** Daylight Factor (0–4 %), mean DF = 0.2 %, DIVA for Rhinoceros & DAYSIM, **b** Daylight Autonomy at 300 lux, mean DA = 10 % of occupied time (8 a.m.–6 p.m., 365 days a year), DIVA for Rhinoceros & DAYSIM

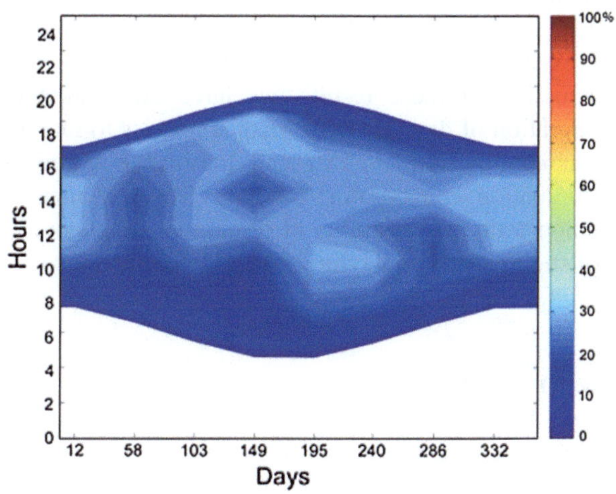

Fig. 6.17 The First Unitarian Church: temporal map of DGP (0–100 %), calculated for each date/time in DIVA for Rhinoceros & DAYSIM, plotted in MATLAB

6.4 Synthesis

Neither the Serpentine Pavilion nor the First Unitarian Church houses programmatic uses that require horizontal task-plane illumination and yet, the majority of our existing daylight design metrics were designed to assess surface illumination and discomfort avoidance for task-related activities. DGP analyzes luminance values within the field of view to predict glare-based discomfort, but this metric was developed for side-lit office environments (Wienhold and Christofferson 2006) and cannot provide relevant data for the programmatic use of a church or glass visitor's pavilion. Both the First Unitarian Church and the Serpentine Pavilion are examples of architecture that suffer under current task and comfort-based daylight performance metrics. Kahn's intent was to create a place of spiritual inquiry and personal reflection. Illumination was used to draw our attention upwards, but it was never designed to accommodate horizontal task-related activities that would necessitate a DF of DA analysis. Annual illuminance variability within the First Unitarian Church, as shown in Fig. 6.14, shows how brightness varies across our field of view. This metric, which is based on the perspective of an occupant, measures perceptual effects within architecture and gives us more relevant information about the visual performance of daylight within Kahn's church. Spatial contrast and luminance variability, as measured within the Serpentine Pavilion, show the designer *where* and *when* the dynamics of contrast and variations in brightness impact our perspective of space.

This chapter demonstrated the successful application of annual spatial contrast and luminance variability metrics to a pair of existing architectural spaces which represent opposing ends of the contrast and variability spectrum. The Serpentine Pavilion produced high instantaneous and annual levels of spatial contrast and luminance variability, while the First Unitarian Church showed moderate-to-low spatial contrast with fluctuating levels of luminance variability throughout the year. Each space was then analyzed for DF, DA, and DGP to compare the results of existing daylight metrics with those proposed by this research. Through a discussion of these results, we have established the need for a more holistic integration of daylight performance metrics to integrate perceptual indicators such as annual spatial contrast and luminance variability with existing task-based illumination and comfort metrics. The new metrics proposed by this research demonstrate the variability of perceptual effects over time and contribute to a more holistic analysis of daylight in architectural space.

References

BS8206-2. (2008). British standard: Lighting for buildings—Part 2: Code of practice for daylighting. National House Building Council.

France, J. (2011). Louis Kahn's First Unitarian Church. Retrieved from http://www.rochesterunitarian.org/Building_desc.html.

http://www.diva-for-rhino.com. (2009). Retrieved from DIVA-for-Rhino.

http://www.rhino3d.com. (2007). Retrieved 2010, from Rhinoceros.

Nabil, A., & Mardaljevic, J. (2006). The useful daylight illuminance paradigm: A replacement for daylight factors. *Energy and Buildings, 38*(7), 905–913.

Reinhart, C., & Walkenhorst, O. (2001). Validation of dynamic radiance-based daylight simulations for a test office with external blinds. *Energy and Buildings, 33*(7), 683–697.

Wienhold, J., & Christofferson, J. (2006). Evaluation methods and development of a new glare prediction model. *Energy and Buildings, 38*(7), 743–757.

Chapter 7
Conclusion

Keywords Spatial contrast · Annual spatial contrast · Annual luminance variability · Integrated architectural design · Daylight analysis · Daylight simulation framework · Lightsolve

7.1 Research Achievements

This book began with a critical look at existing daylight performance metrics through the lens of contemporary architecture. It established the need for new performance criteria that could account for the range of perceptual and temporal qualities within designed space and position those criteria alongside existing task-based illuminance and visual comfort metrics to provide a more holistic analysis of daylight in architecture. Through a survey of global contemporary architecture, we developed a matrix of contrast-driven 'types' to understand the range of daylighting strategies that exist within the language of design. These daylighting strategies reveal a diversity of perceptual effects that rely on spatial and temporal variability within the visual field. Using this typological study as context, we established three new metrics that intuitively describe the perceptual effects within each space. These metrics, spatial contrast, annual spatial contrast, and annual luminance variability, measure the spatial and temporal diversity of daylight in architecture throughout the year.

In order to assign a number to the intuitive effects described above, digital images were used to map luminance levels within a given view, providing a range of values that could be analyzed. Although spatial contrast looks at the variation between neighboring luminance values within a selected image, annual spatial contrast and luminance variability account for the dynamics of contrast and variations in brightness throughout the year. Through an analysis of annual images (56 renderings that represent an even subdivision of hourly and seasonal instances), the designer can identify the magnitude of spatial contrast and luminance variability over time and visualize these dynamic effects through a combination of accumulative spatial images and annual temporal maps. When applied to the ten

S. Rockcastle and M. Andersen, *Annual Dynamics of Daylight Variability and Contrast*, SpringerBriefs in Computer Science,
DOI: 10.1007/978-1-4471-5233-0_7, © The Author(s) 2013

case study spaces in Chap. 5, each metric was successful in producing a linear trend, demonstrating their ability to distinguish contrast and temporal variation across the annual images. The existing architectural spaces also produced convincing results, showing the magnitude of annual spatial contrast and luminance variability within each view. When compared to existing daylight performance metrics such as daylight factor, daylight autonomy, and daylight glare probability, these new annual image-based metrics provide important quantitative information about the dynamic perceptual effects of daylight that have not been previously explored.

These new annual metrics communicate information about the spatial and temporal quality of daylight, giving architects a tool for comparing the magnitude of visual effects within architecture. The implications of this work are widespread, from a simple analytical tool for describing dynamic daylight conditions to an objective approach that challenges the use of task-based illumination and visual comfort metrics in a variety of programmatic conditions. By establishing an intuitive gradient of visual effects and producing a method for quantifying those effects over time, we are able to re-focus the discussion on daylight performance to include those perceptual qualities of light that are often disregarded in contemporary practice.

7.2 Future Research

This research raises an important set of issues for architects and daylight designers. How do we leverage perceptual performance indicators against those task-based illumination and visual comfort metrics that dominate the field of daylight performance analysis? The application of these new metrics to the 10 case studies introduced in Chap. 5 was an attempt to pre-validate their effectiveness in quantifying perceptual effects that were intuitively identified in Chap. 3. To further develop and validate these new image-based metrics, we have identified two avenues of future research. The first path will expand the set of case studies to include a more extensive range of detailed architectural models. This will generate an expanded set of data and allow us to develop a more statistically accurate scale for comparing each metric. The second path will utilize occupant surveys of digital images to validate the relationship between human perceptions of contrast and temporal variability and the proposed metrics.

Other areas of future work include a refinement of the quantitative methods for calculating spatial contrast. The current method takes an accumulative difference between neighboring pixels to produce a set of boundary conditions within a given image. Although this accounts for a fine level of detail in luminance variation (and is very much dependent on pixel density), it does not account for larger areas of contrast that are perceived on a macroscale. A look into contemporary methods of computation, pattern recognition, and cognitive science may produce a multi-scale approach to quantifying spatial contrast.

Ultimately, it is important to propose the integration of these metrics into a software package so that perceptual performance may be measured alongside task-based illumination and visual comfort metrics to provide a more holistic evaluation of daylit space. The Lightsolve project, created at MIT and currently under development at EPFL, proposes an adaptation of these metrics alongside non-visual and dynamic comfort metrics as part of an integrated tool to assess human needs in daylight architecture (Andersen et al. 2013). Through an integration-task-based metric for illumination and visual comfort, photobiological metrics for health, and perceptual field-of-view metrics like spatial contrast and luminance variability, the designer can fine-tune their analysis to fit individualized performance criteria specific to climate, architectural program, and design intent.

Reference

Andersen, M., Guillemin, A., Amundadottie, M., & Rockcastle, S. (2013). *Beyond illumination: An interactive simulation framework for non-visual and perceptual aspects of daylight performance*. Chambery: IBPSA.

MIX
Papier aus verantwortungsvollen Quellen
Paper from responsible sources
FSC® C105338

If you have any concerns about our products,
you can contact us on
ProductSafety@springernature.com

In case Publisher is established outside the EU,
the EU authorized representative is:
Springer Nature Customer Service Center GmbH
Europaplatz 3, 69115 Heidelberg, Germany

Printed by Libri Plureos GmbH
in Hamburg, Germany